Chemical Engineers in Medicine 2015

Topical Conference at the 2015 AIChE Annual Meeting

Salt Lake City, Utah, USA
8-13 November 2015

ISBN: 978-1-5108-1851-4

AA003286

Printed from e-media with permission by:

Curran Associates, Inc.
57 Morehouse Lane
Red Hook, NY 12571

Some format issues inherent in the e-media version may also appear in this print version.

Copyright© (2015) by AIChE
All rights reserved.

Printed by Curran Associates, Inc. (2016)

For permission requests, please contact AIChE
at the address below.

AIChE
120 Wall Street, FL 23
New York, NY 10005-4020

Phone: (800) 242-4363
Fax: (203) 775-5177

www.aiche.org

Additional copies of this publication are available from:

Curran Associates, Inc.
57 Morehouse Lane
Red Hook, NY 12571 USA
Phone: 845-758-0400
Fax: 845-758-2634
Email: curran@proceedings.com
Web: www.proceedings.com

TABLE OF CONTENTS

437503 Mapping Gastrointestinal Disease

Abstract:

The convergence of engineering and medicine presents a compelling opportunity to apply traditional engineering principles to diagnose and treat challenging diseases. For example, the disease eosinophilic esophagitis (EoE) causes frequent ER visits when the esophagus narrows or strictures, blocking flow of saliva and swallowed solids as small as a pea. Patients with EoE may choke without medical help. Evaluation of esophageal tissues in EoE patients finds patchy invasion of eosinophils, a type of white blood cell normally absent. Although endoscopy and biopsy collection is currently recommended for diagnosis, this procedure is invasive, and up to 20% of patients may be misdiagnosed due to the patchiness of eosinophil invasion. Our team is developing a new class of solid ultrasound contrast agents—a slurry of insulin particles—to provide a less invasive way to diagnose EoE and other gastrointestinal (GI) diseases. Patients simply drink the slurry, the particles bind to eosinophil-enriched tissues, and an external ultrasound probe detects the bound particle distribution as a signal of eosinophil concentrations throughout the esophagus. However, significant engineering and clinical questions remain. Can significant surface binding occur even though the bolus transport time is only 6 to 15 s? What liquid rheology most enhances particle binding? What insight can be gained from the recent treatment successes of viscous budesonide, a suspension of glucocordicoid (steroid) particles? How does stricturing affect esophageal fluid mechanics and agent delivery? Can the adult esophagus be imaged through the rib cage? Can these contrast agents be shaped and distinguished in the multiphase milieu of the GI tract? This talk addresses each of these questions essential to the development of this new class of contrast agents to diagnose a broad range of gastrointestinal diseases.

Copyright © American Institute of Chemical Engineers. All rights reserved.

437508 Flow-Induced Segregation in Multicomponent Medical Suspensions

Abstract:

Blood is a suspension of particles of various shapes, sizes and mechanical properties and the distribution of these particles during blood flow is important in many contexts. Red blood cells (RBCs) tend to migrate toward the center of a blood vessel, leaving a so-called cell-free layer at the vessel wall, while white blood cells (WBCs) and platelets are preferentially found near the walls, a phenomenon called margination that is critical for the physiological responses of inflammation and hemostasis. Additionally, drug delivery particles in the bloodstream will also undergo segregation phenomena – the influence of these phenomena on the efficacy of such particles is unknown.

This talk describes efforts to gain a systematic understanding of flow-induced segregation phenomena in blood and other complex mixtures, using a combination of theory and direct simulations of flowing suspensions. A kinetic theory model based on pair collisions and wall-induced hydrodynamic migration can capture the key effects observed in direct simulations, including a "drainage transition" in which one component is completely depleted from the bulk of the flow. Experiments performed in the laboratory of Wilbur Lam indicate the physiological and clinical importance of these observations. More generally, having in hand an understanding of the mechanisms underlying these phenomena now allows more rational approaches to development of quantitative models of them and processes that exploit them. This knowledge will also lead to a better understanding of the consequences of these phenomena in physiology and medicine.

Copyright © American Institute of Chemical Engineers. All rights reserved.

437559 Nanolipoprotein Particles: Encapsulated or Surface-Bound for Biomedical Applications

Abstract:

Nanolipoprotein Particles (NLPs) are disc-shaped nanometer-sized lipid bilayer patches stabilized by a belt of scaffold proteins. NLPs have an average thickness of 5 nm, with a diameter ranging from 10-25 nm depending on the stoichiometric ratios and types of lipids and scaffold proteins being used. This allows NLPs to be compatible with the pore size (5-50 nm) of mesoporous silica. Therefore, we perform entrapment of NLPs using a quick, simple sol-gel processing technique for TMOS that includes evaporation of the majority of the methanol after the hydrolysis reactions. To ensure proper functioning of silica sol-gel entrapped NLPs, we have investigated the phase behavior of the lipids in addition to the secondary structure, localization, and environmental polarity of the scaffold proteins. Our results indicate that silica gel-entrapped NLPs remain intact, with only slightly altered lipid and scaffold protein structure and dynamics. We will briefly discuss the potential to entrap NLPs containing embedded integral membrane proteins (IMPs) for various applications such as biosensing, affinity chromatography, high-throughput drug screening, and bio-reaction engineering.

Further, scaffold proteins can bear polyhistadine tags, which are capable of chelating to Cu2+ metal ions. Lipid-phase specific, iminodiacetic acid (IDA) functionalized lipids are also capable of chelating Cu2+, providing a mechanism for phase-targeted binding of NLPs. We investigate this binding via fluorescence microscopy and characterize interaction with phase-separated supported bilayers and giant unilamellar vesicles (GUVs). The thermodynamics (enthalpy of lipid mixing and steric pressure of protein crowding) and morphology of binding are also examined. Targeted binding of NLPs bearing functional IMPs and/or other biomolecules to supported lipid bilayers has a variety of applications, including development of nano-array technologies and biosensors.

Copyright © American Institute of Chemical Engineers. All rights reserved.

437506 Drug Discovery and Development in a Network of Innovation

Abstract:

Earlier this decade, a respected and widely cited study from Tufts University's Center for the Study of Drug Development came up with the astounding figure of $1.2 Billion as the average cost of bringing novel medicinal drugs to commercialization from the earliest R&D stages of discovery biology. Understandably, this large figure was met with shock and disbelief. Shortly thereafter, several business journals embarked on the investigative task of verification of the calculation and data. What they uncovered was even more surprising. Actual R&D spending over a 15-year period (to smooth out annual fluctuations) divided by the number of new drug approvals indicated company-specific estimates in the range of $3.7 Billion (best) to $11.8 Billion (worst) per drug for a cohort consisting of a dozen major companies representing the pharmaceutical industry. The journalists concluded that even after taking allowance that these figures are crude estimates, the pharmaceutical industry had an unsustainable R&D model.

In a dynamic, free-market based society, inefficiencies on such a large scale attract innovators with disruptive models and the pharmaceutical R&D ecosystem is no exception. An analysis of the pharma industry's R&D workflow suggests that the stage between discovery biology and clinical development, known as (preclinical) lead optimization chemistry, is a major bottleneck. There is now ample evidence that R&D innovation in lead-optimization chemistry can reduce significantly the above cited development cost per drug and this presentation will analyze (via information in the public domain) the modified R&D workflows that incorporate a network of innovation. This development is important to chemical engineers as well as the AICHE by association because discovery research in big pharma is the domain of those trained in the "standard" fields of the biomedical sciences whereas the entities in the network of innovation are embracing the integrative and multiscale talents of chemical engineering graduates.

Copyright © American Institute of Chemical Engineers. All rights reserved.

437512 Mucus Penetrating Nanoparticles: From Concept to Success in the Clinic

Abstract:

The controlled delivery of bioactive molecules to target tissues can significantly improve drug effectiveness while reducing side effects by concentrating medicine at selected sites in the body. Mucus layers coat and protect nearly all entry points into the body that are not coated by skin. Until recently, human mucus was thought to be nearly impenetrable to drug delivery particles even as small as 59 nm in diameter. Particles that become trapped in mucus are typically rapidly cleared from the organ of interest, usually within minutes to a few hours. Thus, while the barrier properties of mucus provide outstanding protection against infection and other potentially toxic substances, they have also thwarted efforts to achieve uniform and sustained drug and gene delivery to mucosal surfaces. This talk will focus on our work to understand the length-scale dependent and adhesion-mediated barrier properties of mucosal fluids, and how this knowledge has guided the development of polymeric nanoparticulate carriers capable of improved drug and gene delivery to the respiratory tract, female reproductive tract, gastrointestinal tract, surface of the eye, and other mucosal tissues.

Copyright © American Institute of Chemical Engineers. All rights reserved.

Growing bone through microsphere scaffolds

Ornella Tempo
McNair Scholar
Rowe Scholar

November 9th, 2015

UCONN

Who cares?

- Every year, more than 500,000 to 600,000 bone graft procedures are performed in the United States to address orthopedic injuries [1,2]
- In 2001 more than 3 million bone graft procedures were performed globally, with estimates of $3.7 billion through 2017 [3,2]

UCONN | SCHOOL OF ENGINEERING

More than one in four Americans has a musculoskeletal impairment

AAOS

American Academy of Orthopaedic Surgeons

UCONN | SCHOOL OF ENGINEERING

Current Treatments

Autograft

- Require harvesting the bone from the iliac crest which involves an additional procedure

Allografts

- Allografts involve transplanting donor bone tissue; this increases the risk of immunoreactions and transmission of infections[1]

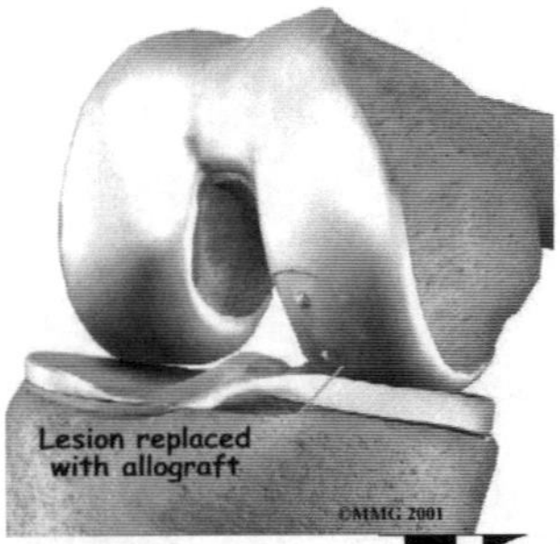

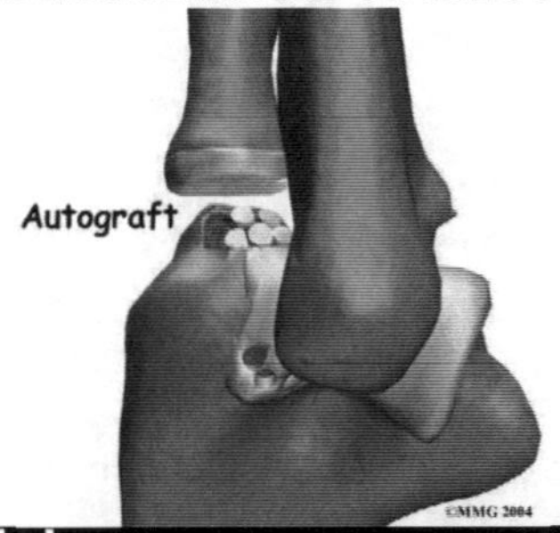

UCONN | SCHOOL OF ENGINEERING

Regeneration of lizards

6-9 weeks for complete regeneration

Have you ever seen the Spiderman movie?

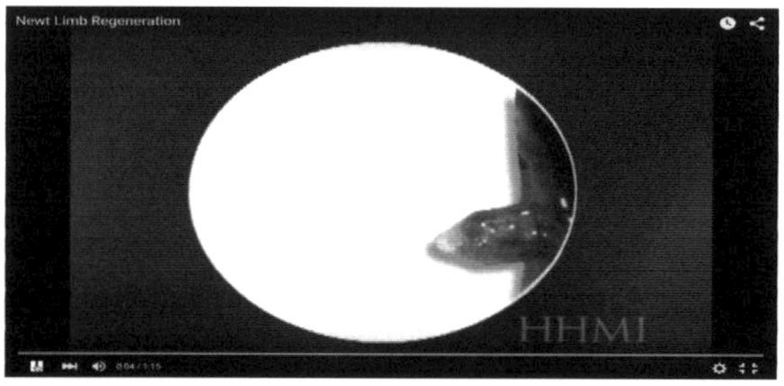

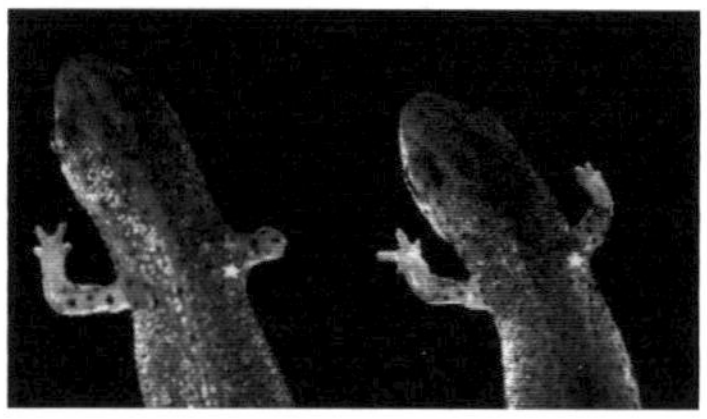

UCONN | SCHOOL OF ENGINEERING

Orthopedic Biomaterials Market 2011-2021

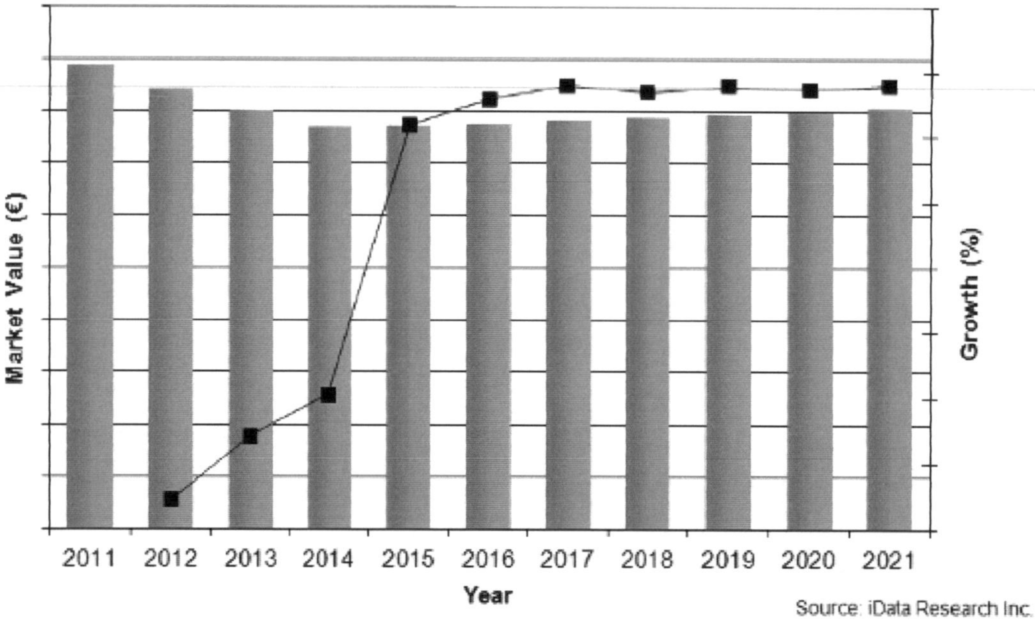

UCONN | SCHOOL OF ENGINEERING

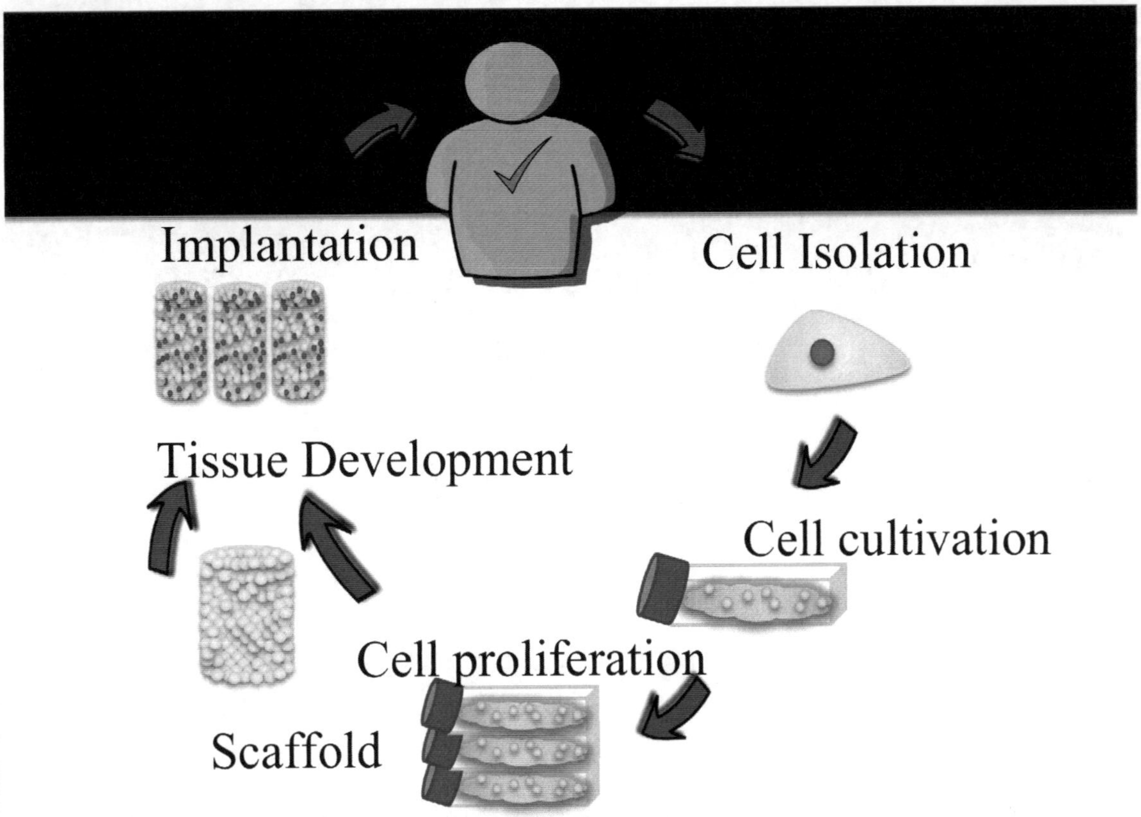

Objective

- To fabricate composite PLGA/CaO_2 sintered microsphere scaffolds as an alternative to current bone repair strategies
- To perform a degradation study to analyze the kinetics and bioactivity of released CaO_2

Hypothesis

- The hydrolytic degradation of composite PLGA/CaO_2 3-dimensional (3D) scaffolds would result in a measurable *in vitro* release of the degradation products such as calcium and hydrogen peroxide over 21 days

UCONN | SCHOOL OF ENGINEERING

Hydrolytic degradation of calcium peroxide

•PLGA scaffolds loaded with calcium peroxide produces H_2O as a degradation product

- $CaO_2 + H_2O \rightarrow CaO + H_2O_2$
- $2H_2O_2 \rightarrow O_2 + 2H_2O$
- $CaO + H_2O \rightarrow Ca(OH)_2$
- $Ca(OH)_2 \rightarrow Ca^{+2} + 2(OH)$

UCONN | SCHOOL OF ENGINEERING

Microsphere Scaffolds

Why microspheres?

- Simple Fabrication
- Control over morphology
- Physicochemical Characteristics
- Versatility of controlling release kinetics of encapsulated factors
- Polymer offers flexibility in degradation kinetics

0.5% CaO_2
25x SEM magnification

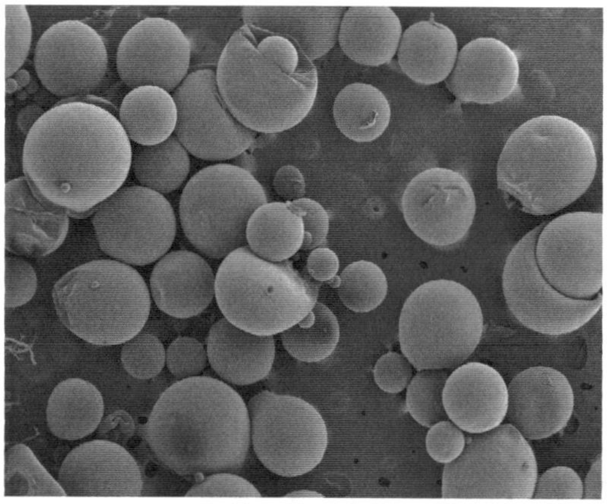

UCONN | SCHOOL OF ENGINEERING

Sintered microsphere scaffold fabrication

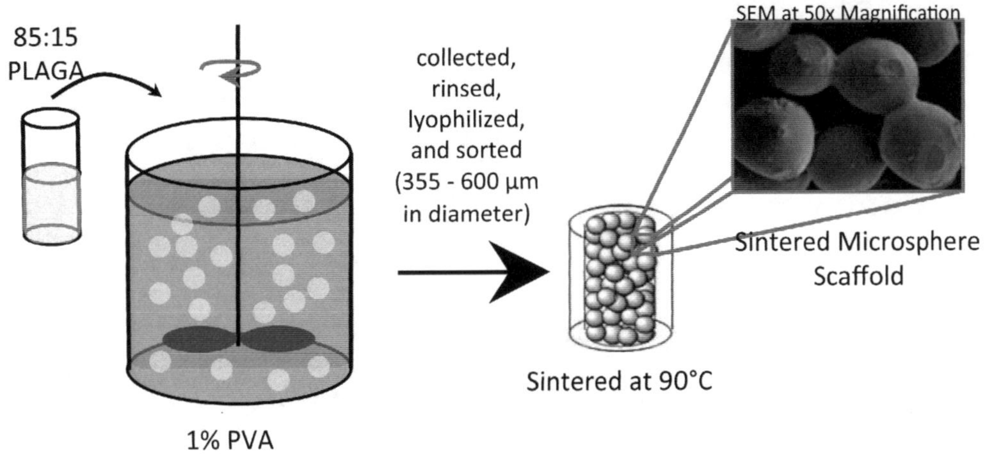

UCONN | SCHOOL OF ENGINEERING

Degradation study

- Calcium Release: The release of Ca^{2+} ions was determined by submerging scaffolds in 1 ml of calcium-free phosphate buffered solution with 500 μL samples taken and replaced at 1, 2, 4, 8 and 12 hours, and 1, 2, 3, 4, 5, 6, 7, 9, 11, 13, 15, 17, 19, and 21 days.

- Quantities of Ca^{2+} were measured using the Calcium (CPC) Liquicolor kit and the hydrogen peroxide was measured with a PeroXOquant Quantitative Peroxide kit

UCONN | SCHOOL OF ENGINEERING

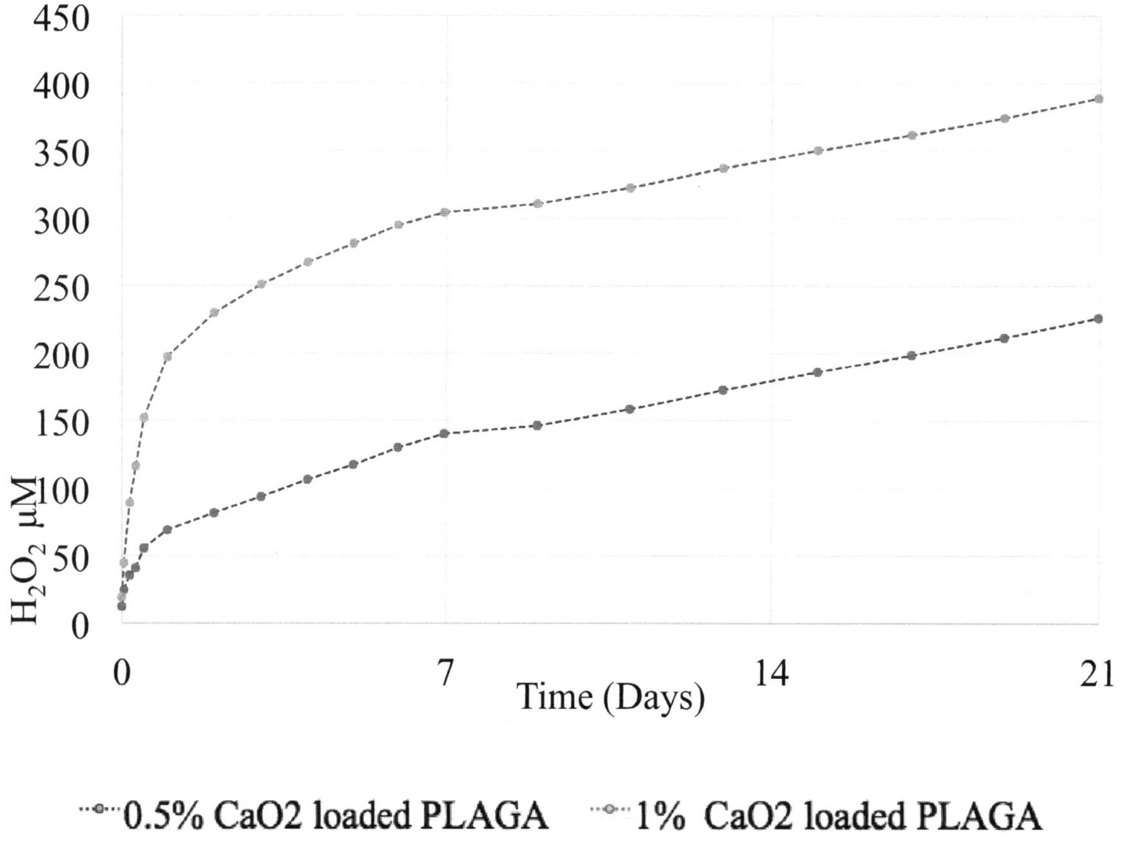

450
400
350
300
250
200
150
100
50
0
H_2O_2 μM
0
7
14
21
Time (Days)
0.5% CaO2 loaded PLAGA
1% CaO2 loaded PLAGA

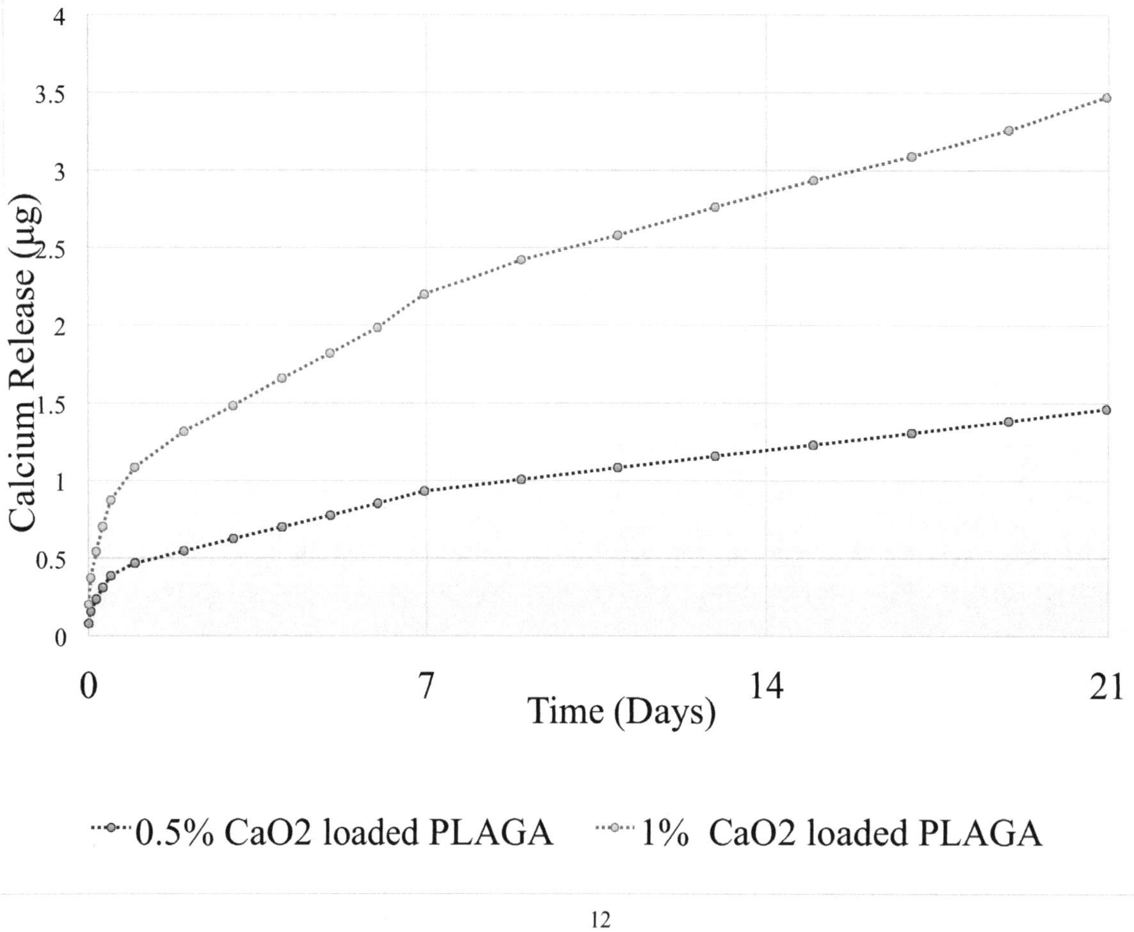

4
3.5
3
2.5
2
1.5
1
0.5
0
Calcium Release (μg)
0
7
14
21
Time (Days)
0.5% CaO2 loaded PLAGA
1% CaO2 loaded PLAGA

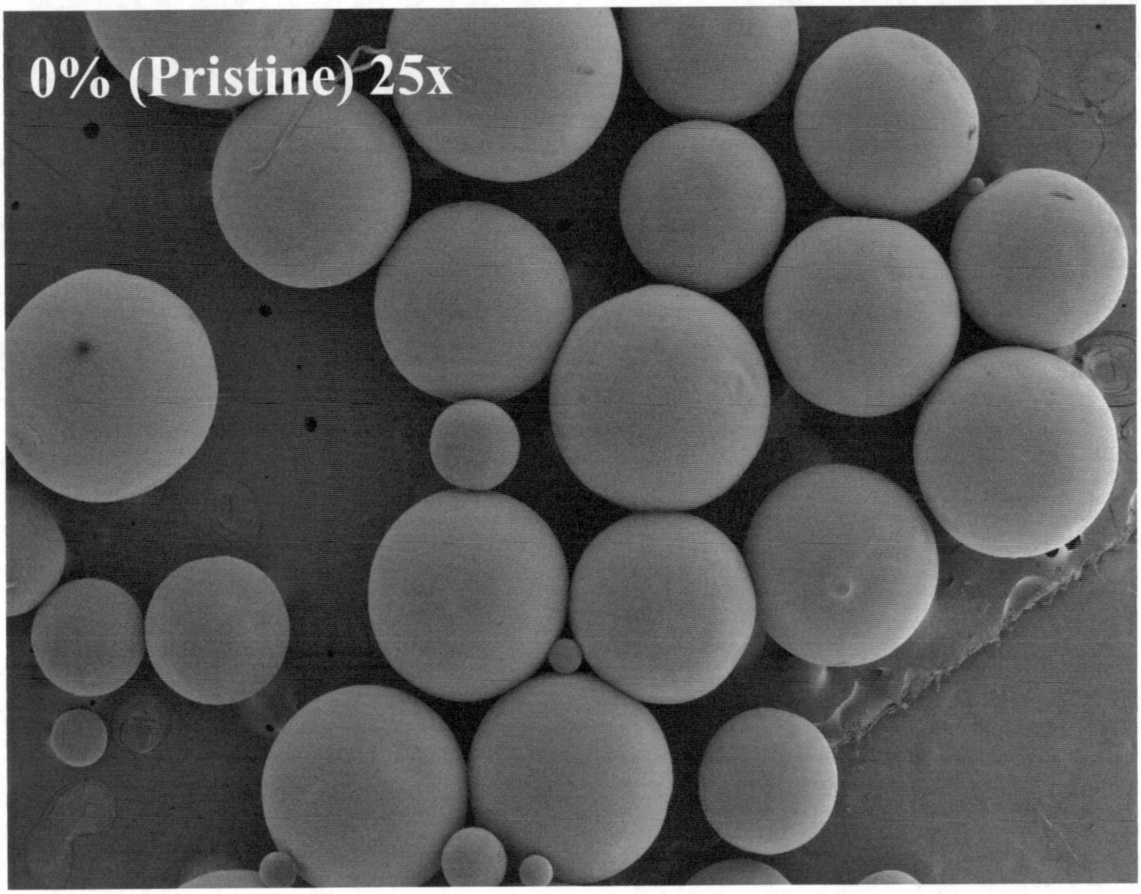
0% (Pristine) 25x

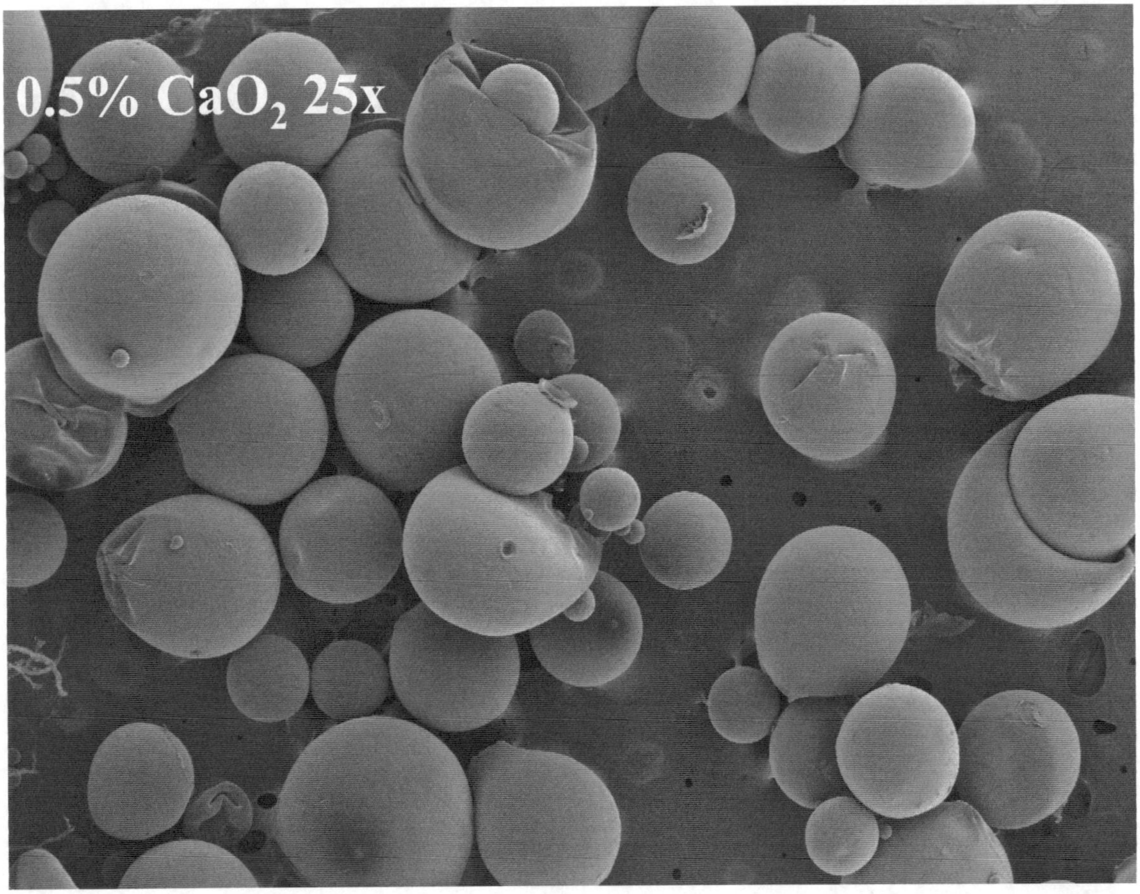
0.5% CaO_2 25x

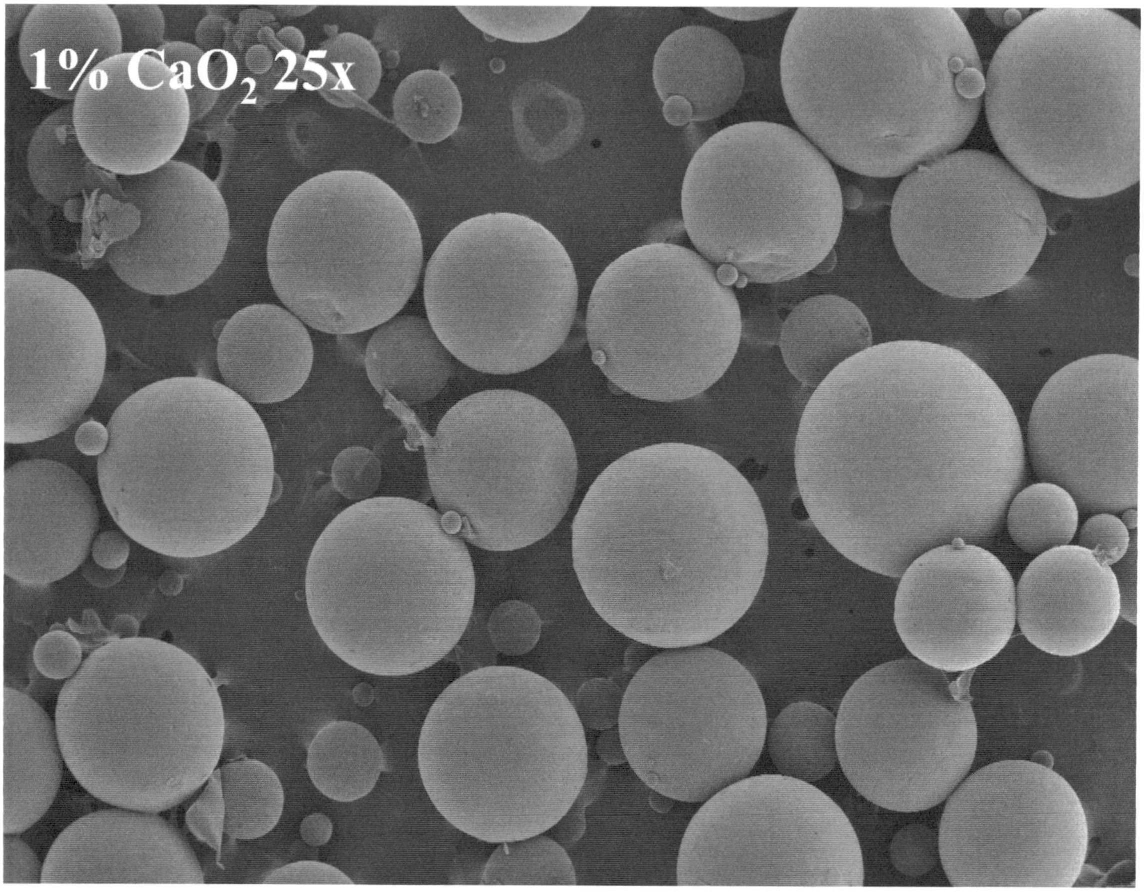

Scaffolds 5mm

Summary

- The *in vitro* release kinetics revealed the structure-function relationship of the material to develop a correlation between *in vitro* and *in vivo* release.
- Calcium levels for low concentration of CaO_2 on the scaffolds increased as time increased. Hydrogen peroxide released levels increases as the concentration of calcium peroxide increases.
- Loading scaffolds with calcium peroxide does not affect its morphology, this is important for cells mobilization, adhesion, proliferation, differentiation and signaling.

- future studies:
  - the observed calcium release could influence the osteogenic differentiation of stem cells without other supplemental factors
  - as noted in recent studies, This new treatment could increase patient access to orthopedic treatments as it is designed to require fewer surgical procedures and no additional technological assistance.

Thank you!

- McNair Scholars Program
- Rowe Scholars Program
- Yusuf Khan, Ph.D.
- Ph.D. Candidate, Keisha Ashe
- Renée Gilberti, Ph.D.

UCONN | CENTER FOR ACADEMIC PROGRAMS

RONALD E. MCNAIR POST-BACCALAUREATE ACHIEVEMENT PROGRAM

UCONN | SCHOOL OF ENGINEERING

419875 Local Delivery of rhBMP-2 from a Compression-Resistant Graft in a Canine Lateral Ridge Augmentation Model

Abstract:

Statement of Purpose: Large mandibular defect reconstruction presents a continual challenge in oral and maxillofacial surgery. Growth factors such as recombinant human bone morphogenetic protein-2 (rhBMP-2) incorporated in scaffolds for tissue engineering promote cellular infiltration, induce osteoblast differentiation, and enhance new bone formation. RhBMP-2 is often delivered via an absorbable collagen sponge (ACS), but this may require a secondary space maintenance system, such as titanium mesh or polymer membrane[1]. Biodegradable polyurethane (PUR) biocomposites are reported to be effective carriers for rhBMP-2 and support new bone growth[2]. Mastergraft ceramic (CM) is an osteoconductive, biphasic ceramic with mineral content similar to bone. PUR/CM composites are osteoconductive and have sufficient stiffness to resist soft tissue prolapse. In the present study, we investigated space maintenance and new bone formation with an injectable PUR/CM composite with and without rhBMP-2 in a canine lateral ridge augmentation model.

Methods: The biodegradable PUR was synthesized from a lysine triisocyanate (LTI) and polyethylene glycol (PEG) prepolymer, a polyester triol (450 g/mol), and triethylene diamine catalyst. Treatment groups included the composite containing 45 wt% CM with no rhBMP-2, a low dose of 200 μg/mL rhBMP-2, or a high dose of 400 μg/mL rhBMP-2 (n=4/group). An ACS with a titanium mesh incorporating 400 μg/mL rhBMP-2 was used as a clinical control. For the PUR composites, lyophilized rhBMP-2 was hand-mixed with the PUR and injected into a lateral ridge defect (2/animal) measuring approximately 13-14 mm mesiodistally, 8-9 mm apico-coronally, and 3-4 mm bucco-lingually. The volume of the injected graft allowed for the expansion to 45-55% porosity to fill the defect space, which occurred within a set time of 7-9 min. After this time, the soft tissue was closed over the defect. For the ACS samples, the rhBMP-2 was resuspended in sterile water and absorbed on the collagen sponge for 15 min prior to placement in the defect. A piece of titanium mesh was fastened over the sponge to prevent soft tissue prolapse. Animals were sacrificed at 16 weeks and new bone formation evaluated by radiographs, μCT, histology, and histomorphometry.

Results: In a previous study we compared PUR/CM and PUR/bioactive glass (BG) composites with a low (100 μg/mL) or high (400 μg/mL) dose of rhBMP-2 in a canine saddle defect model. For both PUR/CM and PUR/BG groups, a high dose of rhBMP-2 led to better preservation of the host ridge width as measured by μCT. PUR/CM composites (a slow-resorbing ceramic) led to improved space maintenance and more new bone than PUR/BG composites (a fast-resorbing ceramic). In the canine lateral ridge augmentation study, the compression-resistant PUR/CM composites are compared to an ACS (clinical control), which requires titanium mesh for space maintenance. To investigate dose dependence on new bone formation, the PUR/CM composite was augmented with 0, 200, or 400 μg/ml rhBMP-2. The canine lateral ridge augmentation study is designed to answer the following questions: (1) Will the biocomposite bone grafts maintain space and prevent prolapse in lateral ridge defects without protective membranes? (2) What is the optimal dose of rhBMP-2 from the PUR/CM composites? (3) How does defect healing compare between PUR/CM composites and the ACS control? In initial surgeries, clinical placement of the PUR/CM composites revealed the ease of conforming to the osseous defects. Once set, the composites maintained their shape during suturing and all sites healed satisfactorily. Preliminary μCT results at 16 weeks showed more new bone formation in high dose PUR/CM composites when compared to composites without rhBMP-2. Work is ongoing to test how the PUR/CM high dose composites compare to the ACS control.

Conclusions: PUR/CM composites with a high dose of rhBMP-2 support new bone formation and remodeling in a canine saddle defect model. We have applied this biocomposite to a canine lateral ridge augmentation model to test the hypothesis that the compression-resistant PUR/CM composite without protective membranes is an effective alternative for rhBMP-2 delivery compared to an ACS control.

Acknowledgements: This work was supported by the Armed Forces Institute of Regenerative Medicine (W81XWH-08-2-0034) and Medtronic, Inc.

References:

1. von Arx T. Int. J. Oral Maxillofac. Surg 2002;31:190-199.

2. Li B. Biomaterials 2009;30:6768-6779.

Copyright © American Institute of Chemical Engineers. All rights reserved.

Nerve Regeneration Using Nerve Growth Factor and Lysophosphatidylcholine

Ryan Wood, Matthew Landeen, Tyler Brown, Austin Thompson, Jonathan Jacobs, Trent Taylor, Stephen Wirthlin, Stephen Wong, Keaton Karlinsey, Mark Rigby, Alonzo Cook

Wallerian Degeneration

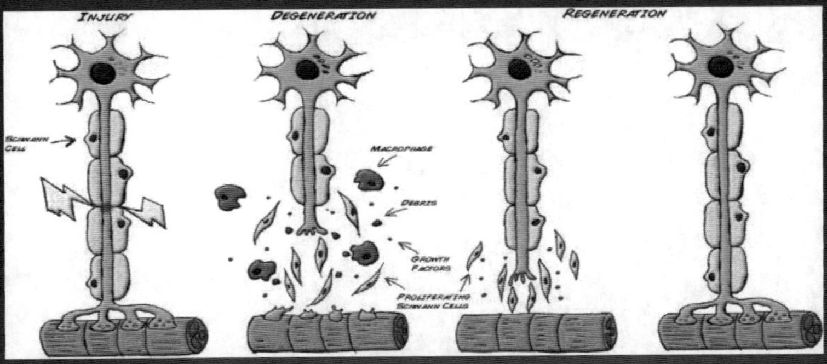

Following injury, Schwann cells detach from the axons, start proliferating and help the recruited macrophages to clear the cellular and myelin debris. At the same time, expression of stimulating factors by Schwann cells create a favorable environment for nerve growth.

Faroni, A. et al. **Peripheral nerve regeneration: Experimental strategies and future perspectives** Advanced Drug Delivery Reviews, Volumes 82–83, 2015, 160–167
http://dx.doi.org/10.1016/j.addr.2014.11.010

Nerve Growth Environment

- During development, neuronal growth is directed by proteins to the target tissue. These proteins include surface-bound matrix molecules, neurotrophins, and growth factors.

- After injury, neurotrophin receptors are upregulated on the distal portion of the nerve, on denervated Schwann cells and on the growth cones of regenerating axons.

- Research has found that Schwann cells near the target tissue could prevent nerve re-innervation after long periods of denervation.

Current Solutions

- Reattachment

- Autograft (Gold Standard)

- Allograft (Not yet widely used)

- Nerve Conduits (Surgical alternative)

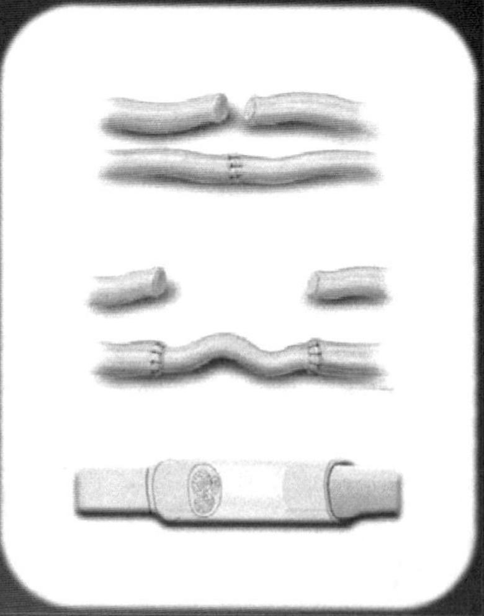

Chemical Treatments

- Some Examples:
  - Neurotrophic Factors
  - Support Cells
  - ECM Proteins

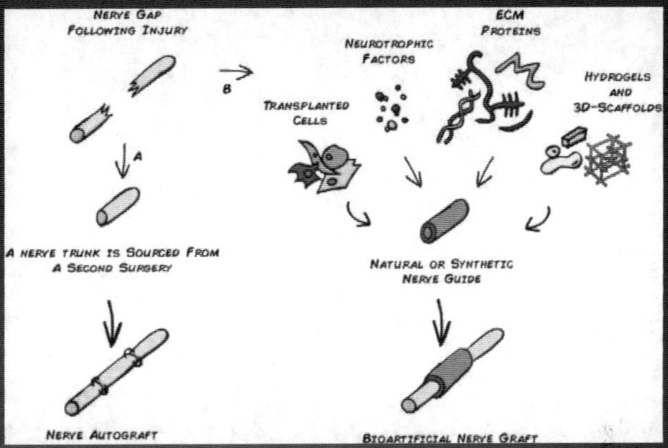

Nerve Growth Factor (NGF)

- NGF is the most characterized neurotrophin and is known to be generated by the cells of target tissues, such as skin, muscle, etc.
- NGF supports survival, growth, and differentiation of neurons in the peripheral nervous system.
- NGF acts through two-types of receptors, the tropomyosin-related kinase A (TrkA) and the p75 receptor for neurotrophins.

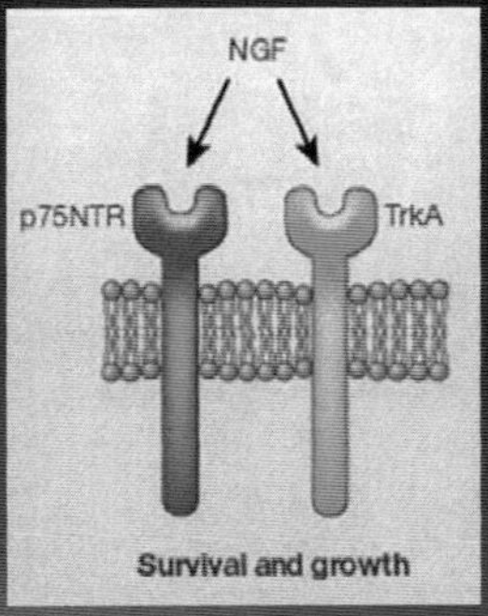

Lysophosphatidylcholine (LPC)

- LPC is a bioactive lipid that gets released from the plasma membrane and the hydrolytic degradation of phosphatidylcholine.
- LPC acts on both the nerves and the on the Schwann cells surrounding the nerves.
- With the nerves, LPC acts through phosphorylation of the Trk-A receptor in the presence of NGF which helps prevent neurite death.
- With Schwann cells, LPC induces prompt demyelination through phospholipase activity and promotes NGF receptor expression.

Proposed Solution

- Lysophosphitydalcholine injections will increase expression of the Nerve Growth Factor receptors which, when coupled with Nerve Growth Factor administration, will result in an increased rate of regeneration.

- 3 main groups: Crush, Crush + NGF, Crush + NGF + LPC

Results

- Electrophysiology

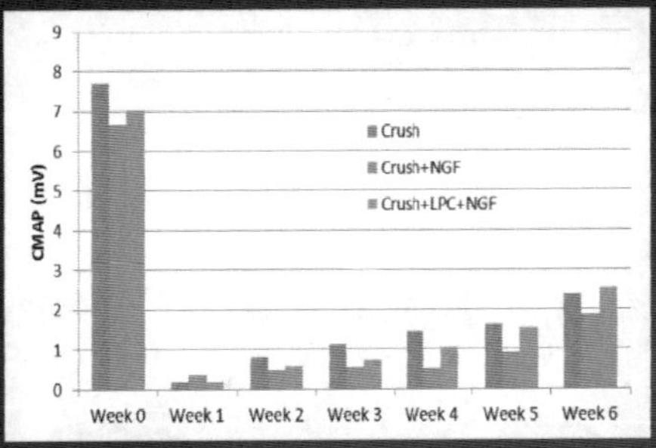

- Histology

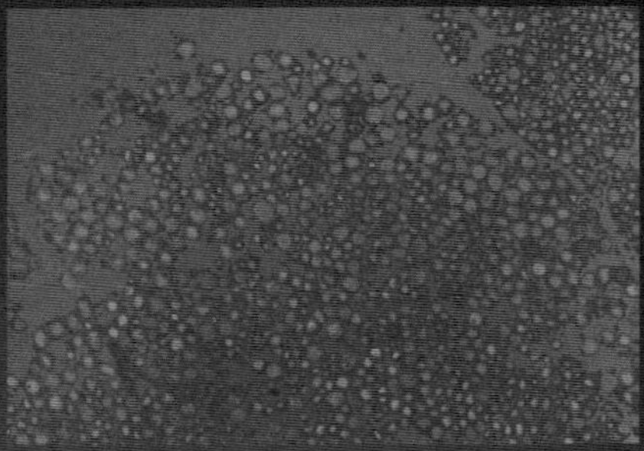

Future Work

- Optimizing NGF and LPC concentrations
- LPC with Multiple NGF injections
- Test healing capacity of just electrical stimulation
- Compare the early stages of demyelination between a crush and a crush receiving LPC

The Lab

- Dr. Alonzo Cook
- Dr. Scott Steffensen
- And all the undergrads!!

- Funding from Brigham Young University

433822 Rational and Combinatorial Analysis of 3D Biomaterials for Optimized Responses of Neural and Non-Neural Cells for Neural Tissue Engineering

Abstract:

Peripheral nervous system (PNS) regeneration after an injury is limited by presence of scar tissue, inhibitory myelin, cell death and lack of growth promotive substrate. Rational design of biomaterials supporting neurons and non-neural cells (Schwann cells and endothelial cells) should lead to an ideal growth promotive nerve guidance channel. We hypothesize rational selection of scaffold, matrix proteins and growth factors and screening of cell responses to combinatorial library will lead to discovery of optimal tissue-specific biomaterial supporting all cell types. To address this goal, we have designed and developed a novel high-throughput screening platform for generation of composite biomaterials and evaluation of cell responses. Composite libraries are generated using a high-throughput printer which combines scaffold and matrix protein at relevant concentrations. Cell encapsulated 3D biomaterials are cultured in microarrays and cell responses are evaluated through high throughput imaging. A confocal flatbed scanner and confocal high content array scanner are used for image acquisition and cell proliferation, morphology and metabolic activity are assessed. We evaluated the cell responses of HUVEC, brain endothelial cells, schwann cells and neurons. Collagen type I was used as base scaffold and matrigel, laminin, fibronectin and collagen type IV were used as matrix proteins. Relevant growth factors to cell type VEGF and FGF (endothelial cells), NGF and BDNF (neurons and schwann cells) are screened. High throughput screening results indicate cell response is dependent upon both collagen and matrix protein concentration and each cell type exhibited unique behavior. The high-throughput screening system provides an efficient way to screen combinatorial biomaterials in a cost-effective, resourceful manner and identify biomaterial 'leads'. The results cell-specific screens will provide useful information for designing optimized tissue-specific biomaterials.

Copyright © American Institute of Chemical Engineers. All rights reserved.

422555 Recellularization of Whole Decellularized Porcine Hearts with Human Cardiac Fibroblasts and Endothelial Cells

Abstract:

Recellularization of Whole Decellularized Porcine Hearts with Human Cardiac Fibroblasts and Endothelial Cells

Nima Momtahan, Beverly L Roeder, Alonzo D Cook

Cardiovascular diseases in the United States account for approximately 30% of all deaths and 60% on a global scale [1]. During myocardial infarction, a portion of the heart does not receive a supply of nutrients and oxygen which leads to irreversible cell damage. For patients who progress to end stage heart disease, the only option is to receive a heart transplant; however, the demand for heart transplants is higher than the supply of suitable hearts. Advances in technology (e.g. total artificial heart, ventricular assist devices, improved diagnostic techniques, etc.) have slowed the accelerating demand for end-stage organ transplantation, yet are not sufficient to satisfy the need. Clinical efforts to restore lost function after an infarction through cell delivery to the injured areas of the heart have not been completely effective [1]. Tissue engineered hearts that are produced by decellularization and then recellularization with the patient's own cells hold promising advantages such as creating an unlimited source of donor hearts [2]. Also tissue engineered hearts are non-immunogenic, meaning they will not be rejected by the recipient's body because the natural scaffold is non-immunogenic, and the cells are sourced from the same patient. Advances made in research of stem cells such as the discovery of iPS (induced pluripotent stem) cells as well as improved decellularization techniques that produce naturally vascularized scaffolds with preserved components, have brought a lot of hope to create the ideal tissue engineered heart.

DNA and some membrane proteins of the cells can initiate an immune response leading to the rejection of the tissue. Ideal decellularization will be achieved by completely removing all cells and cell debris from the heart, while keeping the heart scaffold's vasculature intact and also preserving the important growth factors, glycosaminoglycans (GAGs), and proteins in the extracellular matrix (ECM) [3]. In order to replicate the normal heart's function, the ECM should be reseeded with the necessary cell types and be able to provide support for the cells from preserved structural and signaling molecules. The heart consists of many cell types, including cardiomyocytes, endothelial cells, vascular smooth muscle cells, pericytes, and cardiac fibroblasts in which fibroblasts outnumber all other cell types.

In this work, we improved the decellularization process of porcine hearts in terms of time and cost by using a custom-designed perfusion apparatus. Automation systems were implemented in the system to control pressure and temperature during the process, as well as programmed gradual pressure increase. A combination of type-1 distilled water, phosphate buffered saline, sodium dodecyl sulfate and Triton X-100 with minimum detergent exposure to the ECM were used for the decellularization process. Decellularized samples were characterized by histology and immunohistochemistry staining and tested for residual DNA, GAGs and collagen. Biocompatibility of the decellularized ECM was demonstrated with a hemolysis assay which measured the rupture of erythrocytes in contact with the scaffolds. An immunogenicity test was performed by measuring the stimulation of macrophage cells when they were in contact with the decellularized ECM compared to a fresh heart.

After decellularization, the right ventricles of the hearts were dissected, cannulated through the main right coronary artery and any large open blood vessels were closed using a cyanoacrylate tissue adhesive (Vetbond®). Sections of the hearts were placed in sealed, agitated and oxygenated bioreactors and sterilized with a protocol optimized for being less destructive and enhancing cell attachment to the ECM. Sections were then perfused with growth medium for 24 hours and reseeded with 1×10^8 fluorescence-labeled human umbilical vein endothelial cells (HUVECs) and human cardiac fibroblasts (HCFs). The HUVECs were introduced through vasculature perfusion and HCFs by needle injection into the ECM at multiple time points. After at least 7 days of growth, histological images showed HUVECs dispersed throughout the right ventricle, covering the vasculature and HCFs present in the myocardium.

This study introduced and validated a decellularization method that is optimized for porcine hearts which are similar to human hearts in terms of size and anatomy. This method may be very beneficial in aiding the development of tissue engineered human hearts. We showed that the decellularized ECM has minimal DNA residue and stimulation of macrophages which means that the ECM should not be immunogenic when transplanted into the body. Also the hemolysis assay that was developed showed minimal erythrocyte cell rupture which demonstrates cell compatibility of the decellularized ECM. Furthermore, perfusion recellularization of the right ventricles with HUVECs and HCFs demonstrated the feasibility of recellularizing whole decellularized porcine heart tissues. The next steps will be towards repopulating the ECM with cardiomyocytes in addition to supportive cells, in order to obtain a functional heart tissue that can be transplanted into an infarcted heart to help restore its lost function.

1. Lanza, R., R. Langer, and J.P. Vacanti, *Principles of tissue engineering*. 4th ed. 2014: Academic press. p. 771-789.

2. Robertson, M.J., et al., *Optimizing recellularization of whole decellularized heart extracellular matrix*. PLoS One, 2014. **9**(2): p. e90406.

3. Momtahan, N., et al., *Strategies and Processes to Decellularize and Recellularize Hearts to Generate Functional Organs and Reduce the Risk of Thrombosis*. Tissue Eng Part B Rev, 2015. **21**(1): p. 115-132.

Copyright © American Institute of Chemical Engineers. All rights reserved.

420095 Recellularization of Whole Porcine Kidneys with Human Epithelial and Endothelial Cells

Abstract:

Recellularization of whole porcine kidneys with human epithelial and endothelial cells

Nafiseh Poornejad, Evan M. Buckmiller, Jaron J. Lundwall, Ho Hin Ma, Beverly L. Roeder, Alonzo D. Cook

Brigham Young University, Provo, Utah

Whole organ engineering is a promising alternative for organ transplantation. Murine, porcine and human kidneys have been successfully decellularized and rat kidneys have previously been recellularized with human umbilical vein endothelial cells (HUVECs), rat neonatal kidney cells, and murine stem cells [1, 2]. In this study previous technology was scaled up to porcine kidneys as a model of human-sized kidneys. Decellularization of porcine kidneys was optimized using pressure control and tonic cycles and the effects of freezing/thawing conditions on decellularized extracellular matrix were investigated [3]. Sterilization methods for whole kidneys were also optimized to improve cell adhesion and proliferation. Human renal cortical tubular epithelium (RCTEs) cells and murine endothelial cells (MS-1) were used to repopulate the epithelial and endothelial niches of renal tissue, respectively. More than 6×10^8 endothelial or epithelial cells were perfused through the whole decellularized porcine kidneys for 30 min, either through the ureter for epithelial cells or through the artery for endothelial cells, and then media perfusion was continued for at least 7 days. Immunohistological staining demonstrated that tubules were repopulated with epithelial cells and endothelial cells were located in the complicated vasculature of the kidney. Metabolic activity of reseeded RCTEs was determined by resazurin reduction assay and glucose assay kit. The epithelial cells activity throughout the kidney will be checked by creatinine clearance measurement. This study demonstrated the feasibility of regeneration of whole human-sized kidneys using decellularized porcine kidneys and human cells.

[1] Ross EA et al. Journal of the American Society of Nephrology 20, 2009.

[2] Song JJ et al. Nature Medicine 19,2013.

[3] Poornejad N et al. Organogenesis 11, 2015.

Copyright © American Institute of Chemical Engineers. All rights reserved.

431265 Three-Dimensional Rapid Prototyping of Vascular Substitutes for Medical Applications

Abstract:

Three-Dimensional Rapid Prototyping of Vascular Substitutes for Medical Applications

Connor Dodge, Alex Bischoff, Jason Griggs, Kyle Larsen, Sarah Livingston, Sterling Rosqvist, Tomonori Baba, Tyler Hendricks, Alonzo D. Cook PhD.

Brigham Young University, Provo, Utah, USA

Autologous transplant of a patient's living veins is the most common clinical practice for atherosclerotic disease of the peripheral arteries. Often, it is non-ideal to transect a sick patient's veins to substitute for an artery. The search for an ideal arterial substitute has taken exciting new turns. The newest research focuses on recreating and then implanting arteries from a patient's own cells in vitro, with no immunological effects and having every functional property of a living artery. Our blood vessel research team has entered the tissue engineering field in its most exciting effort: the scalable rendering of cell-seeded vascular constructs with rapid prototyping machines or 3D printers. Gabor Forgacs Ph.D. and other researchers have pioneered the organ printing field. To date, these scientists have managed to build 3-dimensional scaffolds of organs and vasculature while simultaneously seeding living cells using the same machines.

We have built and are modifying a 3D printer that will allow us to test several different avenues of 3D organ genesis. We are currently testing the viability of lyophilized alginate gel, fibronectin coated alginate gel, and methacrylated alginate gel as bio-compatible organ scaffolds. Using these three materials as scaffolds, we hope to learn more about the ideal conditions for cell growth and replication on a polymer scaffold.

We plan to construct blood vessels by 3D printing a cylindrical shape, then preparing it in the correct manner; by either lyophilization, coating with fibronectin, or methacrylation, and attaching the structure to a pump and circulating live endothelial cells through the lumen of the polymer scaffold. We currently perform tests with MS1 endothelial cells, and plan to include smooth muscle cells in our upcoming tests to observe the interaction between endothelial and smooth muscle cells on a polymer scaffold. We expect that with the right pump speed and environment, endothelial cells will attach to the lumen of the polymer scaffold and align themselves as they replicate to cover the inside surface, creating a non-thrombogenic section of blood vessel.

To test functionality we will compare tensile strength with Young's modulus models, test for blood contacting with thrombogenicity measurements in partnership with Thrombodyne Inc., ensure synchronization of smooth muscle contractions, and eventually perform autologous animal tests to identify signs of immunogenic rejection.

Acknowledgements

Funding was provided by Brigham Young University and the Office of Research and Creative Activities (ORCA).

Copyright © American Institute of Chemical Engineers. All rights reserved.

433871 Kinetic Characterization of Proliferation and Dedifferentiation of Müller Cells

Abstract:

Kinetic Characterization of Proliferation and Dedifferentiation of Müller Cells

Alonzo D. Cook, Chris Burns, Ryan Gillis, Steven Christiansen

Approximately 11 million Americans suffer from dry age-related macular degeneration (dry AMD), an eye disease that occurs in the macula of the eye when retinal pigment epithelial (RPE) cells experience oxidative stress and die. In several animals (including rats, zebra fish, and chicks), photoreceptor regeneration is possible with the appropriate stimulus. Müller cells have been shown to be the critical cell in photoreceptor regeneration. In zebra fish and chicks, Müller cells dedifferentiate to a proliferating progenitor cell when photoreceptors are damaged. These progenitor cells are then transported to their final area in retinal tissue where they differentiate to photoreceptors and integrate into the retinal tissue. Activation of cellular signaling pathways (Notch, MAPK, and others) is integral to this process. On each level of regeneration (dedifferentiation, proliferation, differentiation, migration and integration), these signaling pathways guide cell behavior, producing new photoreceptors that replace dead or damaged photoreceptors. A therapy guided by this knowledge could replace more intrusive methods to potentially cure dry AMD.

Dedifferentiation and differentiation of Müller cells is a potential option for curing dry AMD. The therapy would involve inducing Müller cells to transform into the needed photoreceptors. This would likely involve initially dedifferentiating the Müller cell into a progenitor state and then inducing the progenitor cells to become photoreceptors. Dedifferentiation is the process by which a cell that has already developed into a fully functioning adult cell returns back to a previous state. The dedifferentiation step has been qualitatively observed in vitro. Determination of the proliferation kinetics was performed in-house by isolating Müller cells and obtaining time-resolved cell counts. A Wistar albino rat eye was dissected and the retina was removed, dissociated and cultured. The Müller cells were isolated and divided into several different 25 mL flasks. An initial cell concentration was measured with a microfluidic cell counter. At 12 hour time intervals, a flask was removed and cell counts were taken.

Assuming a 1st order model of proliferation dependent on cell concentration, the rate constant was found to be 0.09/hr for cells cultured with fetal bovine serum. This rate constant dropped to 0.03/hr when fetal bovine serum was removed from the growth media. This value matches data reported by Limb[1] and is similar to the value derived from data reported by Zhao[2]. The shape of the cells was observed to shift to a dendritic neuronal shape. The dedifferentiated nature of the cells was confirmed by immunocytochemistry, observed qualitatively in the cells used in the proliferation experiments, and quantitatively described by protein ratios according to the method of Goldman, et al[3]. The concentration of proteins typical of photoreceptor progenitor cells was assumed to be proportional to the number of cells, and the rate constant for dedifferentiation was determined to be 0.35/hr. This is significant because when combined with proliferation kinetic data, it is clear that dedifferentiation occurs much faster than proliferation.

This project will continue by repeating the proliferation kinetic experiments and attempting to determine the rate constants for differentiation into photoreceptors. Also the model will be modified to handle non-first order rates. Eventually, once the system is defined, the model will allow the optimal design of therapies for retinal degeneration.

References

1. Limb, Astrid; et al. Differences between the neurogenic and proliferative abilities of Müller Glia with stem cell characteristics and the ciliary epithelium from the adult human eye. *Exp Eye Res.* 93(6): 852-61. Dec 2011.

2. Zhao. Induction of Retinal Progenitors and Neurons from Mammalian Müller Glia under Defined Conditions. *J. Bio. Chem.* Feb 2014.

3. Goldman, Daniel; et. al. Ascl1a regulates Müller Glia dedifferentiation and retina regeneration via a Lin-28-dependent, let-7 miRNA signaling pathway. *Nat Cell Bio.* 12(11): 1101-7. Nov 2010.

Copyright © American Institute of Chemical Engineers. All rights reserved.

438527 Controlled Release from Polyelectrolyte Complex Drug Carriers

Abstract:

Comparison studies of release of a model drug, Rhodamine B, from two different types of polyelectrolyte complex beads (PECs) made of poly(styrenesulfonate, PSS)-poly(allylamine, PAA) and chitosan-alginate were investigated. The release was carried out by using a modified Franz diffusion cell into deionized water at 37°C. The long term sustained release (> 7 days) of Rhodamine B was obtained with PSS-PAA PECs due to their tighter networks. By tuning the compositions of PSS-PAA, the time needed to completely release the drug could be varied from two days to 14 days. For the case of chitosan-alginate PECs, 100% accumulated release occurred within 24 hours. With the incorporation of a thermo-responsive polymer, e.g., poly(N-isopropylacrylamine), the release could be extended to two or more days for the chitosan-alginate PECs. By drying the beads, the release of Rhodamine B was found to be further extended.

Copyright © American Institute of Chemical Engineers. All rights reserved.

438450 3D Printed Microfluidic Device for Dynamic Investigation of the Blood Brain Barrier

Abstract:

The "Blood Brain Barrier" (BBB), surrounding the capillaries in the Central Nervous System, is a tissue composed of endothelial cells joined to each other by tight junctions. It serves to impede the passage of any impurities from the blood to the brain and is essential to brain homeostasis, disruption of which can result in a myriad of brain disorders like Alzheimer's, Multiple Sclerosis, and Parkinson's disease. Current in-vitro models of the BBB include a seeded Transwell ® Membrane under static conditions, where the apical and basal side fluids do not exhibit dynamic flow and apply shear stress on the membrane. Being a simplified model of the tissue, it does not adequately model the tissue as it is in-vivo, where there is dynamic flow of both basal (cerebrospinal fluid) and apical (blood) fluids. Consequently, we have designed a microfluidics device, the μTRANS chip, to model the BBB under dynamic conditions, allowing us to study not only the tissue structure of the BBB but also the cellular level dynamics of it under different environmental conditions. Our initial results, revolving around the construction of the device, its characterization, and the biochemical/mechanical modulations of the tissue will be presented at the conference. In terms of construction, the device was built using a seeded Transwell ® membrane (consisting of a layer of human brain endothelial cells on the apical side and a layer of astrocytes on the basal side) inserted between two glass slides. Each of the glass slides served as a distinct chamber, containing two 3D printed micro-channels, interconnected by smaller micro-channels. The smaller channels were designed to allow for fluid exchange between the larger micro-channels. Gold electrodes were also built into the device to analyze the permeability of the BBB using electrochemical impedance spectroscopy (EIS). The channels, glass substrate layers, and Transwell ® Membrane were all placed into a 3D printed vacuum chamber as the final step in construction, in order to maintain contact between the different layers of the device during experimental modulations. The permeability characteristics of the BBB were studied using fluorescence and EIS, specifically in the form of transendothelial electrical resistance tests (TEER), under different biochemical and mechanical conditions, including varying levels of pH, shear stress, and nutrients. We believe that our ability to study the BBB in a dynamic environment allows us to not only accurately model the tissue under physiological conditions but to control its characteristics using biochemical and mechanical means.

Copyright © American Institute of Chemical Engineers. All rights reserved.

438448 Self-Assembly Simulations of Polymer Functionalized Virus Capsids

Abstract:

Viral capsids have been the subject of a significant amount of research in a variety of fields. Their potential relevance to biomedical science arises from their ability to encapsulate therapeutic agents and deliver them to targeted cellular sites. The capsids can be decorated with polymers to target specific sites or promote controlled aggregation for creating novel materials. We used coarse-grained molecular dynamics to simulate the aggregation dynamics of amphiphilic functionalized icosahedral capsids, representing the Cowpea Mosaic Virus capsid, when grafted with diblock copolymers of polyethylene glycol and polylactic acid. Additional variations of the system includes changing the polymer grafting density and capsid volume fraction to show how these changes affect the self-assembled structures. Our observations suggest that (a) the excluded volume of the icosahedral and (b) the polymer grafting density play the largest role on aggregation dynamics and morphologies when compared to the other variants.

Copyright © American Institute of Chemical Engineers. All rights reserved.

438410 Investigation of the Interaction Between a Novel Drug Delivery System and an Epithelial Cell Layer

Abstract:

Surgery and other invasive processes are the current treatment options for patients with diabetic retinopathy, glaucoma, and age-related macular degeneration, which are eye diseases that affect at least nine million Americans. We are developing a targeted drug-delivery system, which would provide a non-invasive option for treatment. The system features a collagen membrane with drug-loaded poly(lactic-co-glycolic acid) (PLGA) nanoparticles incorporated into a thin membrane. This membrane can be worn by a patient in conjunction with a contact lens, and provides for continuous drug release. The purpose of this project is to study the interactions of lidocaine-loaded nanoparticles in a collagen membrane with a layer of corneal epithelial cells. Lidocaine-loaded PLGA nanoparticles were made and characterized by measuring size and drug content. Nanoparticles were mixed into a collagen solution and then gelled to form a thin membrane. Corneal epithelial cells were cultured on semipermeable well inserts to form a confluent layer of cells. The cell layer was exposed to the collagen membrane containing the lidocaine-loaded nanoparticles. Measurements were taken to determine cell viability and permeability, as well as the amount of lidocaine that crosses the cell layer. Results showed that lidocaine was released from the system and delivered across the cell layer. The release profile shows an initial burst release, followed by steady drug release over the testing period. Cell viability decreased with the application of the membranes, compared to a control sample of cells only. Further work is being done to determine the cause of cell death and to test the system for longer periods of time

Copyright © American Institute of Chemical Engineers. All rights reserved.

438407 Quantitative Analysis of Fundus Images for Grading of Vitreous Haze

Abstract:

Purpose: Intraocular inflammatory processes lead to progressive accumulation of cells and protein exudate in the vitreous, which can be detected by ophthalmoloscopic inspection of the eye as decreased visibility of the retinal vasculature, optic nerve head, and other fundus details. The objective of this work was to develop a quantitative method for grading the blurriness of ocular fundus images in order to provide automated grading of the severity of uveitis.

Methods: A computer algorithm was developed to quantify image blurriness through application of entropy filtering and Fourier spatial-frequency analysis. The algorithm was refined using a set of 8 reference images that were created by optically filtering a single standard fundus picture to varying degrees that simulate increasing vitreous haze. It was then applied without modification to a dataset of 120 digital fundus images collected from patients with different grades of uveitis. The algorithm performance was evaluated against the acutance method, which is commonly used to quantify the overall sharpness of edges in an image. Computed scores of vitreous haze were also compared for both methods in a masked fashion with the subjective readings of three expert clinicians.

Results: The Pearson correlation coefficient between computed automated grades of the algorithm and clinician grades of blurriness was 0.81-0.88 (R^2 = 0.66-0.78). Acutance measurements calculated from the raw images and from images with background fluctuations subtracted out did not perform as well. The Pearson correlation coefficient between acutance and clinician grades was found to be -0.05 (R = 0.003) and 0.33 (R=0.109), respectively, which is quite weak.

Conclusions: We have developed a computer algorithm for grading vitreous haze in an unbiased and quantitative manner that correlates strongly with the subjective readings of expert clinicians. The algorithm performance with fundus images is much better than that of standard acutance measures.

Copyright © American Institute of Chemical Engineers. All rights reserved.

McCormick
Northwestern Engineering
Department of Chemical and Biological Engineering

Developing a Strategy for Constructing Modular Biosensors

Andrew K.D. Younger[1], **Neil C. Dalvie[2]**, Austin G. Rottinghaus[3], and Joshua N. Leonard[2,4,5,*]

[1] Interdisciplinary Biological Sciences Graduate Program, Northwestern University, Evanston, IL, 60208, USA
[2] Department of Chemical and Biological Engineering, Northwestern University, Evanston, IL, 60208, USA
[3] Department of Chemical and Biological Engineering, University of Wisconsin-Madison, Madison, WI, 53706, USA
[4] Chemistry of Life Processes Institute, Northwestern University, Evanston, IL, 60208, USA
[5] Member, Robert H. Lurie Comprehensive Cancer Center, Northwestern University, Evanston, IL, 60208, USA

1

McCormick
Northwestern Engineering
Department of Chemical and Biological Engineering

Motivation: Quantifying metabolism

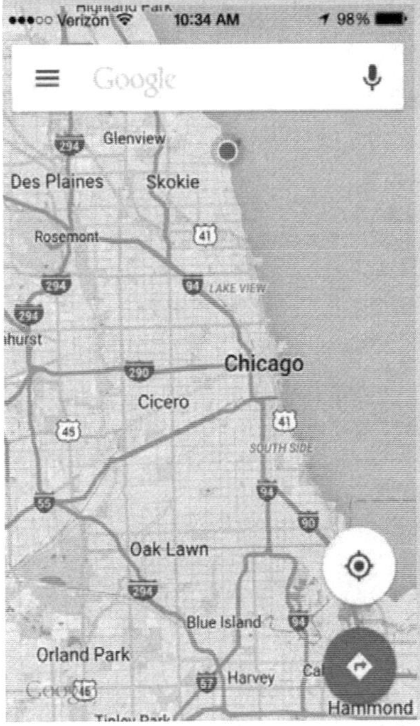

2

McCormick

Northwestern Engineering

Department of Chemical and Biological Engineering

Motivation: Quantifying metabolism

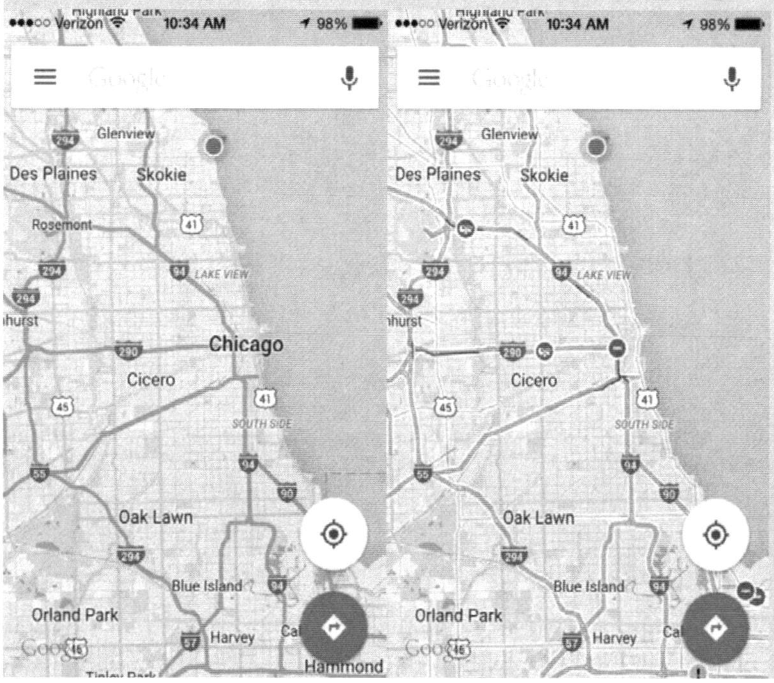

3

McCormick

Northwestern Engineering

Department of Chemical and Biological Engineering

Motivation: Quantifying metabolism

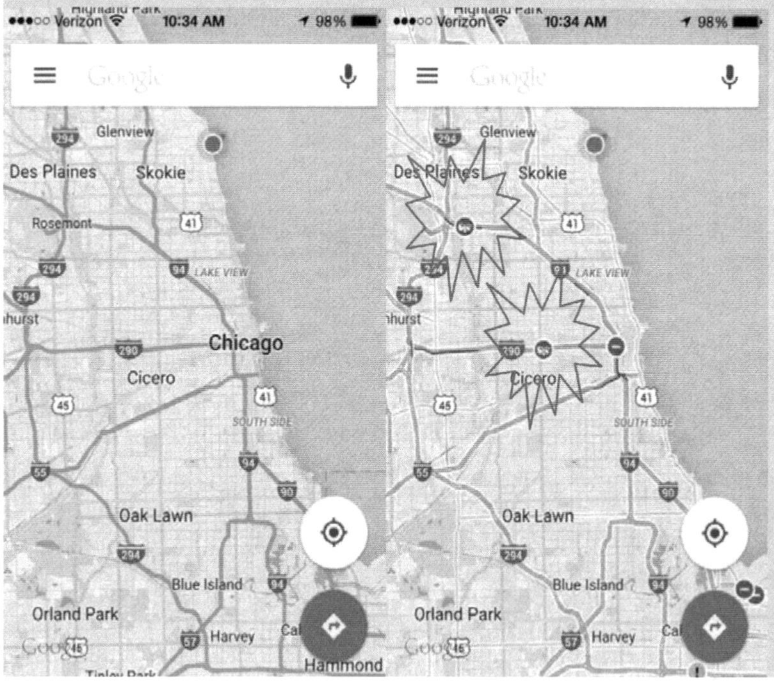

4

McCormick

Northwestern Engineering

Department of Chemical and Biological Engineering

Motivation: Quantifying metabolism

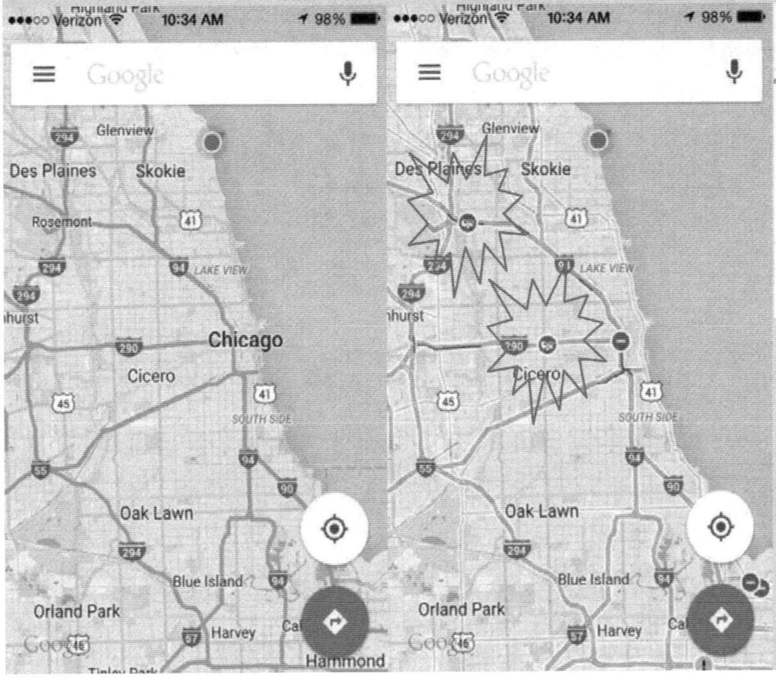

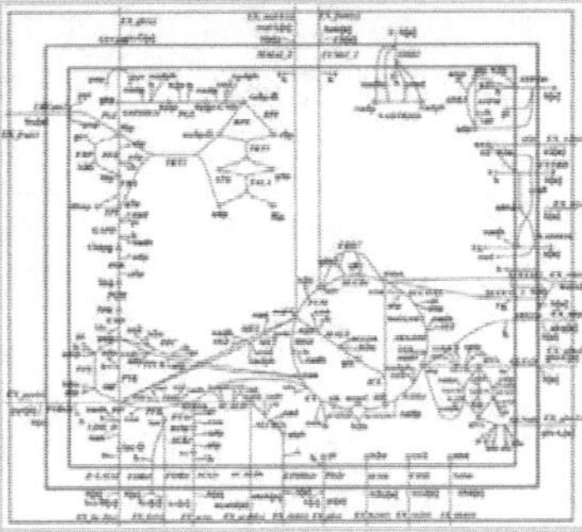

5

Orth *et al.* 2011

McCormick

Northwestern Engineering

Department of Chemical and Biological Engineering

Motivation: Quantifying metabolism

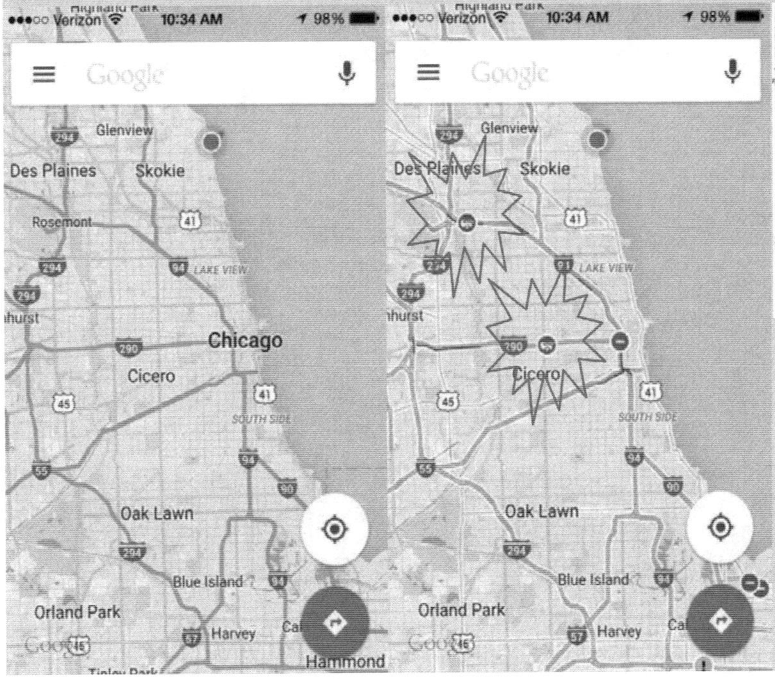

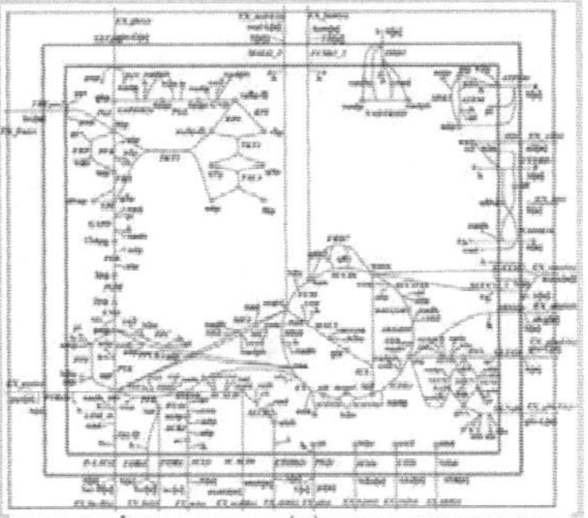

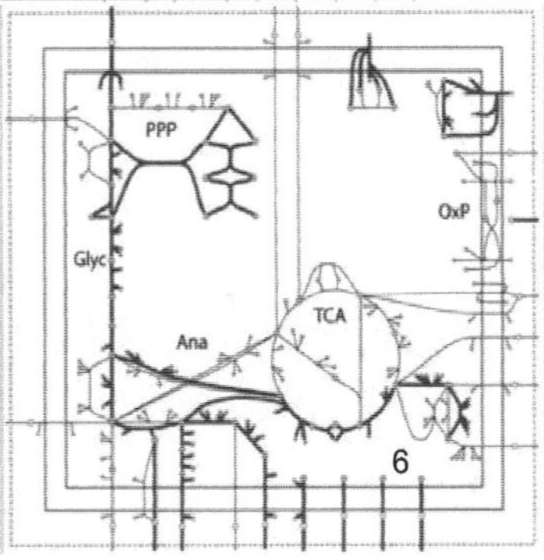

6

McCormick
Northwestern Engineering
Department of Chemical and Biological Engineering

Useful functions of metabolite biosensors

High Throughput Library Screening

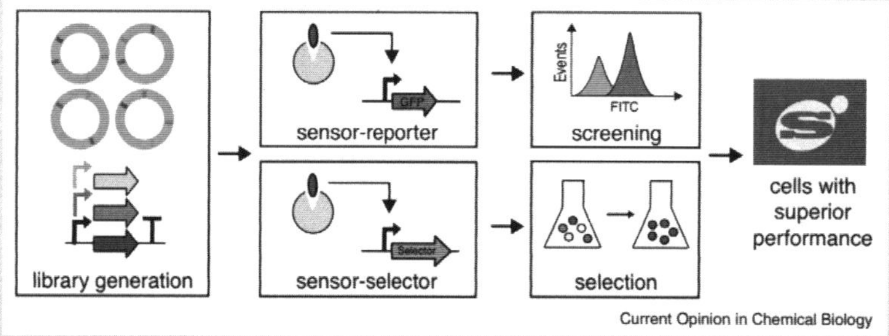

7

Zhang *et al.* 2015

McCormick
Northwestern Engineering
Department of Chemical and Biological Engineering

Useful functions of metabolite biosensors

High Throughput Library Screening

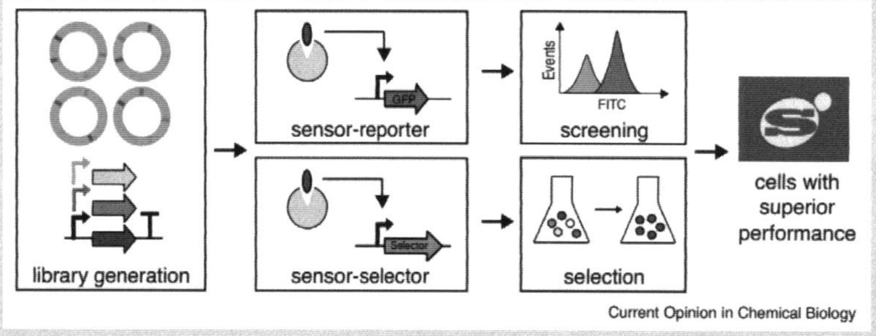

Feedback control

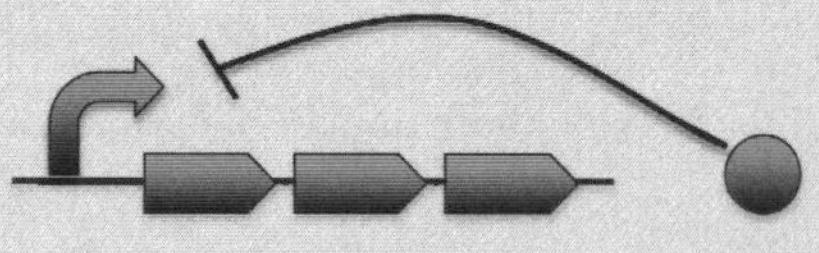

8

Zhang *et al.* 2015

Useful functions of metabolite biosensors

High Throughput Library Screening

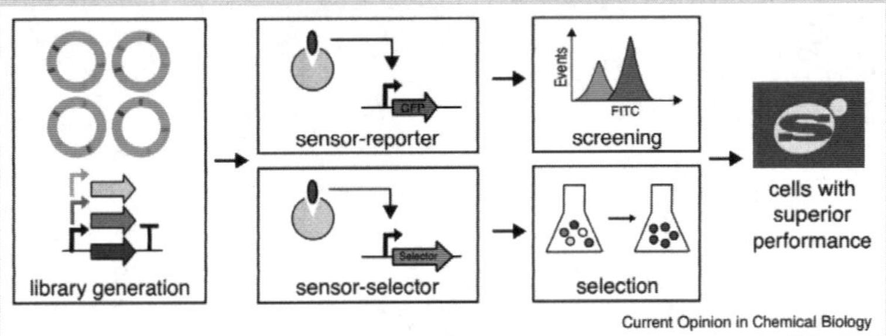

Current Opinion in Chemical Biology

Feedback control

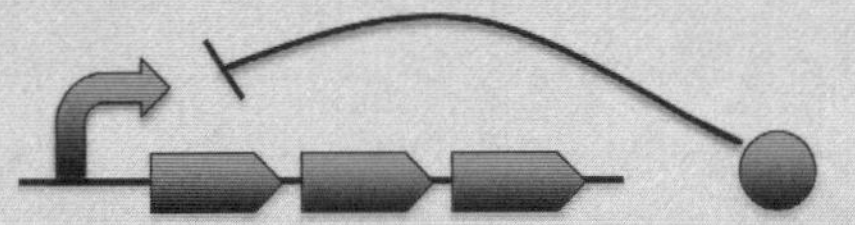

Limited natural options

9

Zhang *et al.* 2015

McCormick
Northwestern Engineering
Department of Chemical and Biological Engineering

Metabolite responsive transcriptional regulators

Goal:

Ligand responsive transcriptional regulator

Approach:

Modular design

-Convert ligand binding protein to biosensor

-DNA binding domain

10

McCormick

Northwestern Engineering

Department of Chemical and Biological Engineering

Metabolite responsive transcriptional regulators

Goal:

Ligand responsive transcriptional regulator

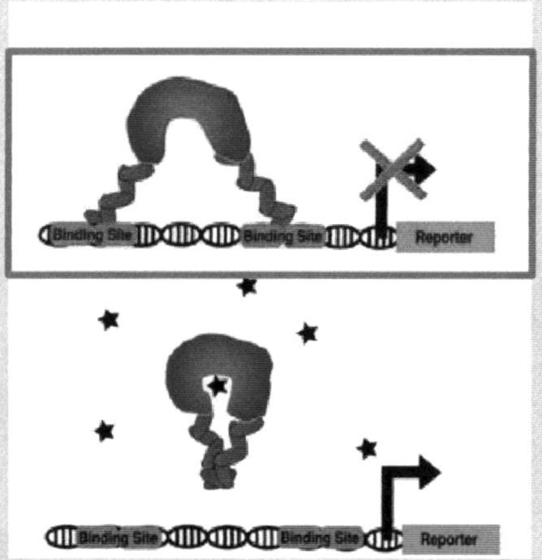

Approach:

Modular design

-Convert ligand binding protein to biosensor

-DNA binding domain

11

McCormick

Northwestern Engineering

Department of Chemical and Biological Engineering

DNA binding domain: BCR1 zinc finger

12

PDB #1A1L Thomas Splettstoesser

McCormick
Northwestern Engineering
Department of Chemical and Biological Engineering

DNA binding domain: BCR1 zinc finger

9 base pair binding site

13

PDB #1A1L Thomas Splettstoesser

McCormick
Northwestern Engineering
Department of Chemical and Biological Engineering

DNA binding domain: BCR1 zinc finger

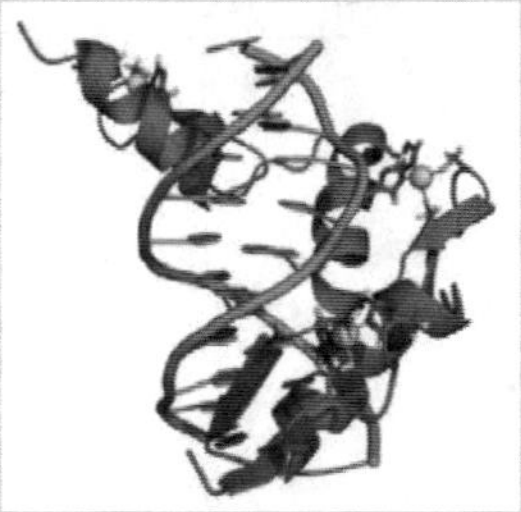

9 base pair binding site

Constitutive GFP reporter

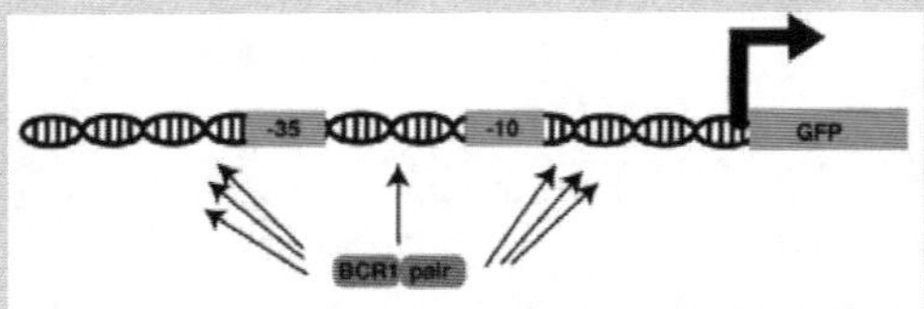

14

PDB #1A1L Thomas Splettstoesser

McCormick

Northwestern Engineering

Department of Chemical and Biological Engineering

DNA binding domain: BCR1 zinc finger

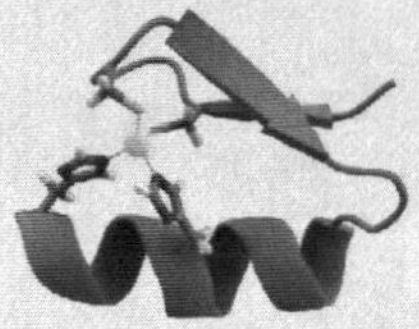

9 base pair binding site

Constitutive GFP reporter

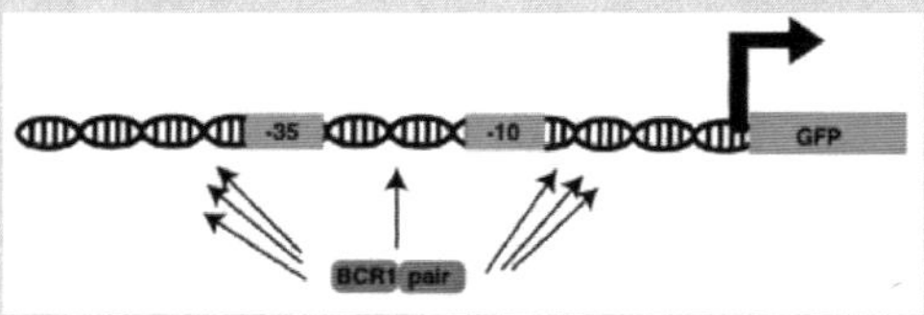

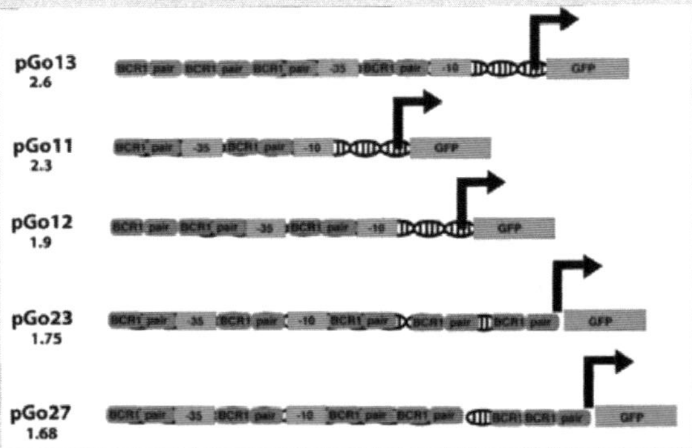

15

PDB #1A1L Thomas Splettstoesser

McCormick

Northwestern Engineering

Department of Chemical and Biological Engineering

Investigate expression of promoter library

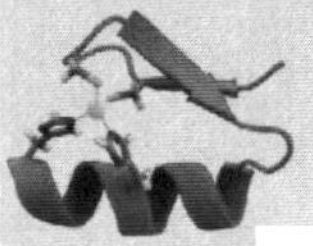

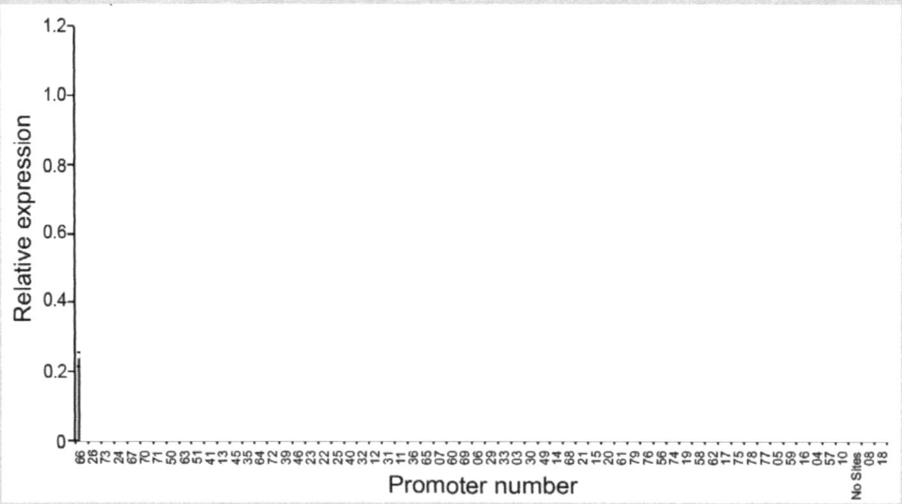

McCormick

Northwestern Engineering

Department of Chemical and Biological Engineering

Investigate expression of promoter library

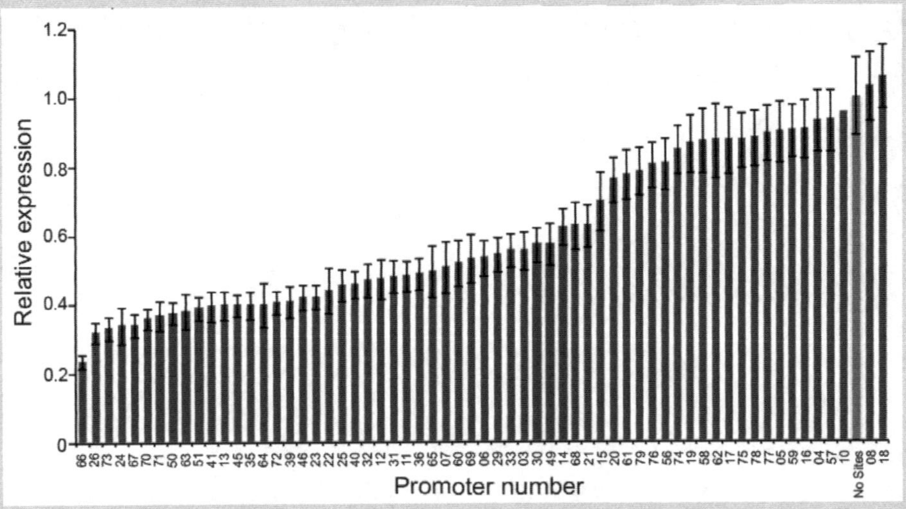

McCormick

Northwestern Engineering

Department of Chemical and Biological Engineering

Investigate expression of promoter library

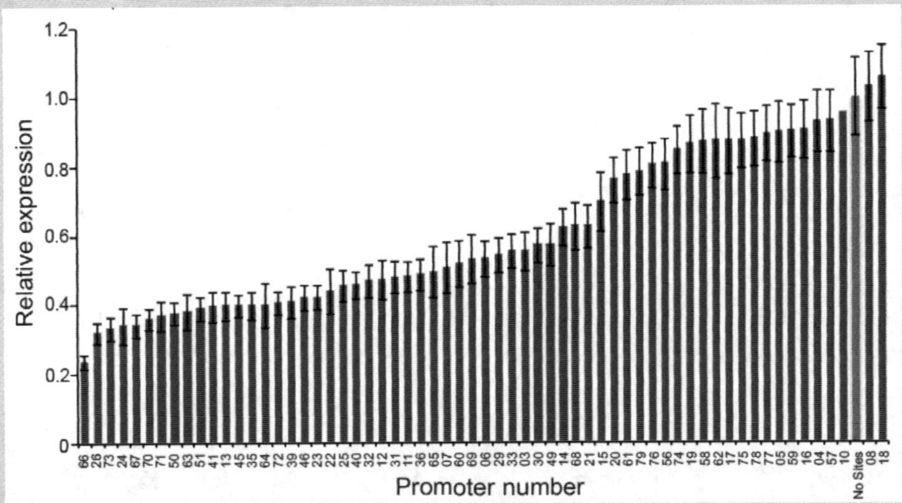

Wide range of repression – Why?

McCormick
Northwestern Engineering
Department of Chemical and Biological Engineering

Method for quantitatively describing the library

Goal: Select a set of descriptors that describe promoters.

Risk: Too many descriptors will over fit the system.

19

McCormick
Northwestern Engineering
Department of Chemical and Biological Engineering

Method for quantitatively describing the library

Goal: Select a set of descriptors that describe promoters.

Risk: Too many descriptors will over fit the system.

Assumption: The only thing that affects reporter expression is the binding of the biosensor

Features only describe the location of zinc finger binding sites

20

McCormick
Northwestern Engineering
Department of Chemical and Biological Engineering

Number	Shorthand	Description of Feature
1 ★	*BS in core*	Number of binding sites in the core
2 ★	*pairs in core*	Number of pairs in the core
3	*BS upstream*	Number of binding sites upstream of the -35 region
4	*pairs upstream*	Number of pairs upstream of the -35 region
5	*BS downstream*	Number of binding sites downstream of the -10 box
6	*pairs downstream*	Number of pairs downstream of the -10 box
7	*length*	Length of the promoter in base pairs
8 ★	*spacer -10*	Base pairs downstream of the -10 box, before the first site
9 ★	*spacer -35*	Base pairs upstream of the -35 region, before the first site
10 ★	*spacer inside -10*	Base pairs upstream of the -10 box, before the first site
11 ★	*spacer insider -35*	Base pairs downstream of the -35 region, before the first site
12	*core middle space*	Non-binding site base pairs in the core
13	*ave downstream spacing*	Average number of base pairs between sites downstream
14	*ave upstream spacing*	Average number of base pairs between sites upstream
15	*down length*	Base pairs downstream of the -10 box
16	*up length*	base pairs upstream of the -35 box
17	*lopsidedness*	Absolute value of (sites upstream) - (sites downstream)

21

McCormick
Northwestern Engineering
Department of Chemical and Biological Engineering

Determining important features

Define output:
"repressibility" = (-1)*expression

Questions:

What is the relative importance of features in determining repressibility?

How many of these features are actually significant?

22

McCormick

Northwestern Engineering

Department of Chemical and Biological Engineering

Approach: Three independent feature selection methods

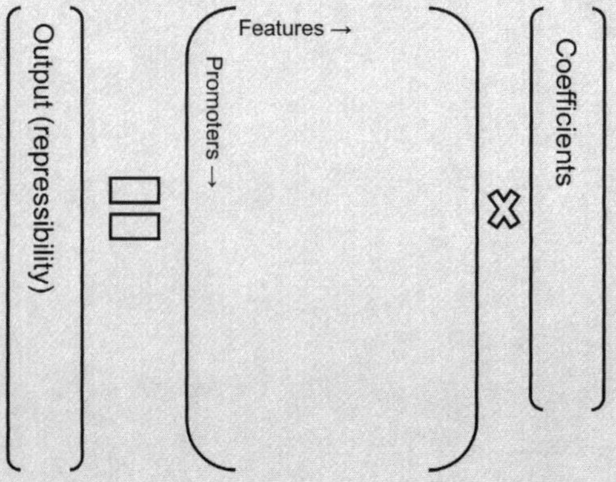

- Partial least squares regression
- Random forest
- Lasso regression

Iterative

Regularized

23

McCormick

Northwestern Engineering

Department of Chemical and Biological Engineering

How many features are significant?

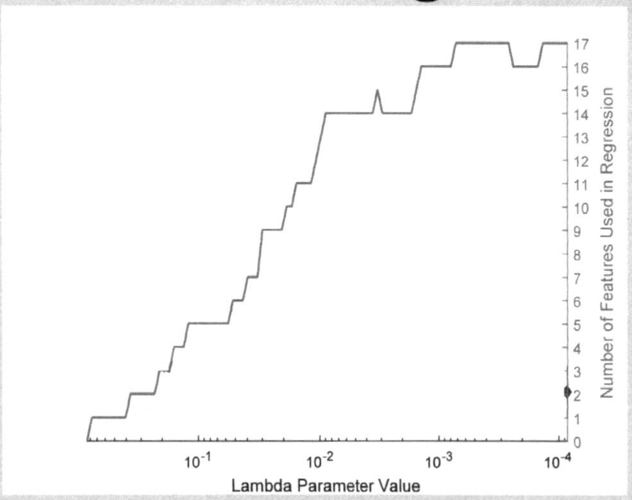

Decreasing penalty on coefficients

Lasso regression shrinks the coefficients of unimportant features to 0

After a certain number of features, the fits stops improving

24

McCormick

Northwestern Engineering

Department of Chemical and Biological Engineering

How many features are significant?

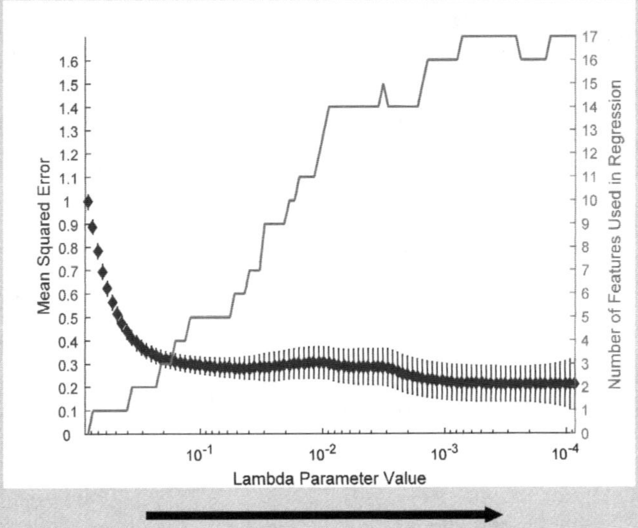

Decreasing penalty on coefficients

Lasso regression shrinks the coefficients of unimportant features to 0

After a certain number of features, the fits stops improving

25

McCormick

Northwestern Engineering

Department of Chemical and Biological Engineering

How many features are significant?

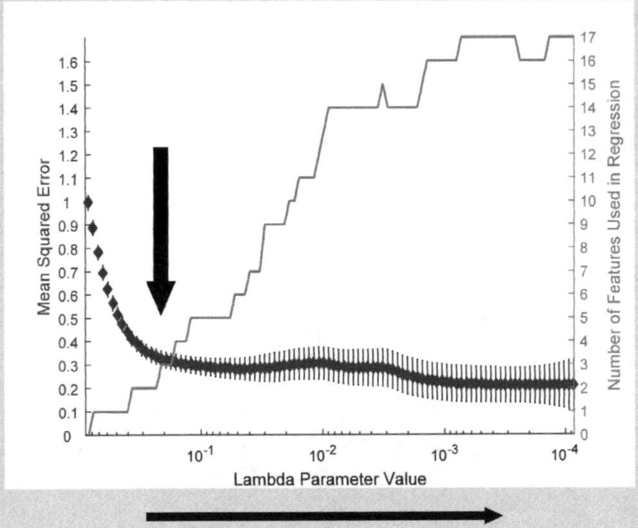

Decreasing penalty on coefficients

Lasso regression shrinks the coefficients of unimportant features to 0

After a certain number of features, the fits stops improving

26

McCormick

Northwestern Engineering

Department of Chemical and Biological Engineering

How many features are significant?

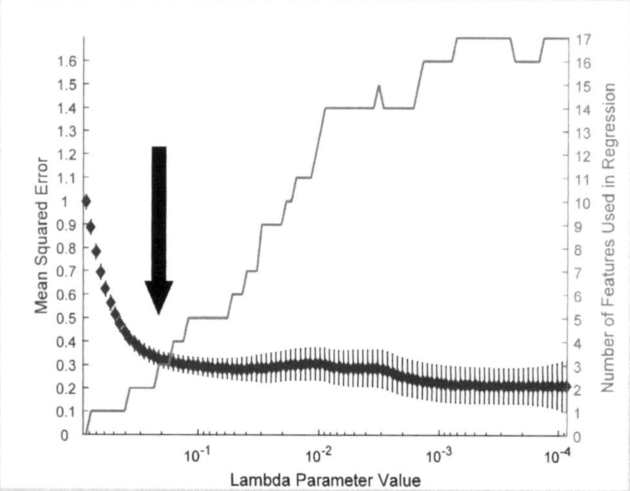

Decreasing penalty on coefficients

Lasso regression shrinks the coefficients of unimportant features to 0

After a certain number of features, the fits stops improving

~3 important features

27

McCormick

Northwestern Engineering

Department of Chemical and Biological Engineering

Comparison of all three methods

Green = positive regression coefficient (positively correlated with repressibility)

Pink = negative regression coefficient (negatively correlated with repressibility)

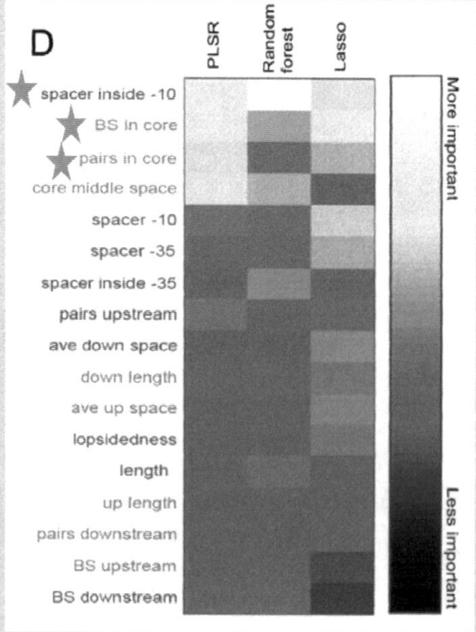

28

McCormick
Northwestern Engineering
Department of Chemical and Biological Engineering

Rules for zinc finger repression

Crowding promoter boxes, especially -10, tends to create strong repression

Binding site quantity does not matter
- No titration effect

Regions away from the boxes are not important (upstream, downstream)

Proposed mechanism: The zinc finger blocks the sigma factor when near the -10 box

29

McCormick
Northwestern Engineering
Department of Chemical and Biological Engineering

Application to biosensors

Promoter library analysis revealed elegant design rules for synthetic promoters

Useful feature selection methods for describing transcription factor binding

Application of this DNA binding domain in biosensors

McCormick
Northwestern Engineering
Department of Chemical and Biological Engineering

Application to biosensors

Promoter library analysis revealed elegant design rules for synthetic promoters

Useful feature selection methods for describing transcription factor binding

Application of this DNA binding domain in biosensors

Development of a small molecule responsive transcriptional regulator platform.
Andrew Younger

Session 734: Emerging Tools for Synthetic Biology

Thursday 3:15, Salt Palace Convention Center 150G

McCormick
Northwestern Engineering
Department of Chemical and Biological Engineering

Acknowledgements

Josh Leonard
Andrew Younger
Austin Rottinghaus
Rachel Dudek
Andy Scarpelli
Michelle Hung
Kelly Schwarz
Taylor Dolberg
Alia Zander
Danny Wells

Collaborators
Declan Bates (U. Warrick)
Paul Freemont (Imperial College London)
Jay Keasling (UC Berkeley)
Anne Osbourn (John Innes Center)
Susan Rosser (U. Edinburgh)

Chemistry of Life Processes Institute

Office of Undergraduate Research UR@NU

McCormick

Northwestern Engineering

Department of Chemical and Biological Engineering

McCormick

Northwestern Engineering

Department of Chemical and Biological Engineering

Decision tree regression: Random forest

Uses decision trees on subsets of features

Quantify how much accuracy is lost when feature is not present

Creates a ranked list of importance

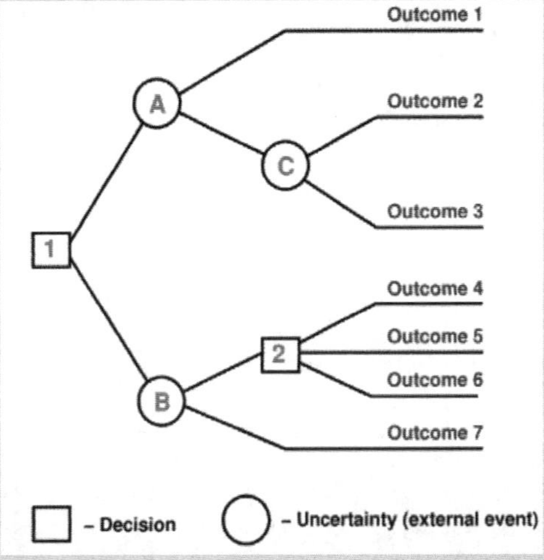

34

McCormick

Northwestern Engineering

Department of Chemical and Biological Engineering

Partial Least Squares Regression: Permutation Test

Shuffle output randomly

Obtain a variance for each feature coefficient

Compare the real output to the random variance

Again, creates a ranked list

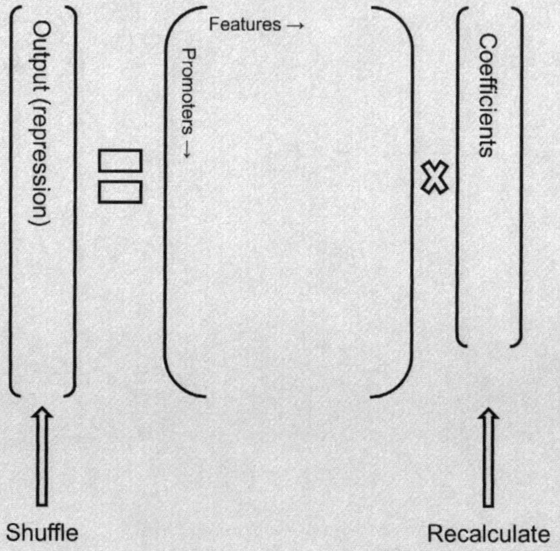

35

McCormick

Northwestern Engineering

Department of Chemical and Biological Engineering

Regularized regression: Lasso

Constrains $\sum|\beta|$
(sum of absolute coefficients)

penalty

$$\sum_{i=1}^{N}(y_i - \sum_j \beta_j x_{ij})^2 + \lambda \sum_j |\beta_j|$$

Eliminates features by shrinking unimportant coefficients to 0

Creates a ranked list of importance (again)

36

438528 Immunomodulatory Amphiphilic Polyanhydride Microparticles for Peripheral Nerve Regeneration

Abstract:

: Peripheral nerve repair using implanted biomaterial conduits seeded with cells face many challenges, including the host immune response. Tissue engineering approaches often employ the use of drug delivery vehicles to supply sustained release of trophic factors which support the differentiation and continued growth of the implanted cells. However, host immune cells responding to inflammatory cytokines released by the damaged nerve may inhibit this growth. This project focuses on the fabrication of amphiphilic degradable polyanhydride micro particles that can release surface adsorbed Interleukin-4 and encapsulated nerve growth factor simultaneously. Interleukin-4, an anti-inflammatory cytokine, serves to promote an M2-like phenotype in invading macrophages as a method to support wound healing. These M2-like macrophages may reduce the inflammatory response, thereby generating an environment which supports cell growth and tissue regeneration, thus allowing implanted cells to benefit from the sustained release of neurotrophic factors. The release of nerve growth factor helps stimulate the peripheral nerve regeneration. We have focused on the fabrication and release kinetics of the two macromolecules from the polyanhydride microparticles.

Copyright © American Institute of Chemical Engineers. All rights reserved.

435067 Nanostructured Vapor Deposited Surface Treatments Improve Bone-Anchored Hearing Aid Integration

Monday, November 9, 2015: 3:15 PM
250A (Salt Palace Convention Center)

Michelle Stolzoff[1], Jason Burns[2], Arash Aslani[3], Eric Tobin[3] and Thomas J. Webster[4], (1)Bioengineering, Northeastern University, Boston, MA, (2)N2 Biomedical, Bedord, MA, (3)N2 Biomedical, Bedford, MA, (4)Chemical Engineering, Northeastern University, Boston, MA

Introduction: Loss of one's sense of hearing can be very detrimental to a person's quality of life with significant limitation in communication. 360 million people in the world have disabling hearing loss, with a rate of 347 per 100,000. Bone-anchored hearing aids (BAHA) are bone-conduction-based hearing aids with titanium implants embedded in the skull. Up to 17% of patients with percutaneous BAHAs have adverse skin reactions around the implant despite strict hygiene practices and antibiotic regimens. This is particularly problematic in children, whose speech/language development is in a critical stage, there are failure rates up to 37%, with 25% requiring explantation. To address this, we investigated the use of nano-featured surfacing for these implants to both reduce infection and inflammatory responses while improving osseointegration. Nanomaterials have been found to have a profound effect upon cell-material interactions, including prevention of bacterial proliferation as well as heightened mammalian cell growth for tissue regeneration. With varied deposition parameters we have demonstrated topographical control of these surfaces and observed changes in bacterial and osteoblast adhesion and integration.

Materials and Methods: Uniform titanium thin films were deposited onto the surfaces of titanium coupons. Changes in vapor deposition parameters allowed us to obtain nano-featured coatings and control the surface topography and extent of nano-roughening. This in turn allowed us to modify surface energy and resulting cellular interactions. Bacterial studies were performed with *S. aureus, S. epidermidis* and *P. aeruginosa*, and seeded at approximately 20,000 bacteria per sample. After 24 hours of incubation, bacteria were removed and counted using a colony-forming unit (CFU) assay as well as a live/dead assay. For osteoblast adhesion and proliferation studies, hFOB human osteoblasts (ATCC) were seeded at 10,000 cells/mm^2 and quantified via MTS after 4 hrs and 1, 4 and 7 days. All experiments were completed in triplicate and repeated at least three different times.

Results and Discussion: Treated Ti coupons had clear changes in visible appearance and in nano-scale topography. When imaged by atomic force microscopy (AFM), 100-nm scale peaks were formed after treatment (Fig 1) with a greater roughness (Rq = 23 nm, 59 nm) as ion beam intensity increased than untreated Ti (Rq = 4 nm). In addition to increasing measured roughness, the topography of the samples could be tuned for peak morphology (spiky vs rounded) as well as frequency (larger singular peaks surrounded by smaller features or evenly-distributed peaks at high concentrations). Ti coated surfaces had much less adherent bacteria (*S. aureus*) after 24 than plain Ti, supporting past observations with other nano-rough materials.

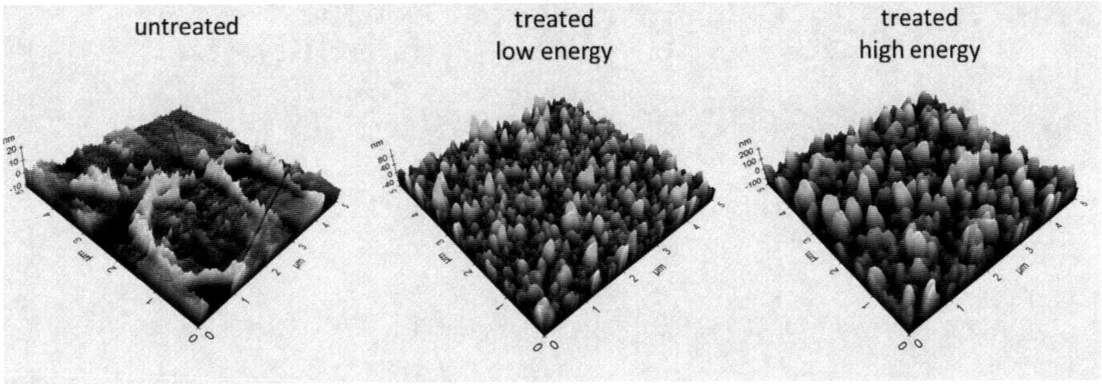

Figure 1. Atomic force microscopy of untreated (left), and samples treated with low and high (middle and right) ion beam energy.

Conclusions: Vapor deposited titanium coatings on Ti can impart controllable nano-scale texturing. Here, we demonstrated antibacterial and pro-osteoblast properties of the resulting nano-rough surfaces.

Acknowledgements: The authors thank Northeastern University and the College of Engineering for funding.

Effects of Red Allotrope Selenium Nanoparticles on Head and Neck Squamous Cell Carcinoma

Christopher Hassan[1] and Thomas J. Webster[1,2]
[1]Department of Chemical Engineering, Northeastern University, Boston, MA, USA 02215
[2]Center of Excellence for Advance Materials Research, King Abdulaziz University, Jeddah, Saudi Arabia
10/9/2015

1

Background on Head and Neck Squamous Cell Carcinoma (HNSCC)

- Stats
  - 500,000+ new cases per year.
  - 50% morality rate.
  - 5-year life maximum expectancy at best.
- Risk factors
  - Typically found in middle-aged smokers who drink heavily.
  - Also found to develop in young women with no apparent risk factors.
    - HPV causally linked
- Onset
  - Loss of chromosomal region 9p21
- Frequent metastasis
  - Lymph nodes

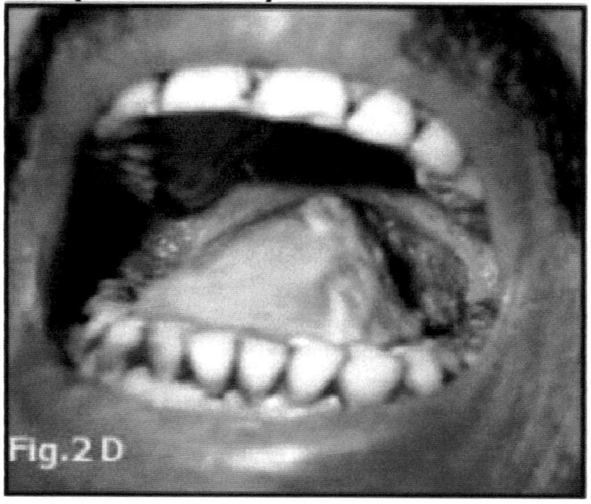

http://openi.nlm.nih.gov/imgs/512/6/2912870/2912870_1471-2407-10-369-2.png

- Perez-Ordonez, B., M. Beauchemin, and R. C. K. Jordan. Journal of Clinical Pathology, 2006, 59, 445 – 453.
- Carvalho, Andre Lopes, Ines Nobuko Nishimoto, Joseph A Califano, and Luiz Paulo Kowalski. International Journal of Cancer, 2004, 114: 806 – 816.
- Forastiere, Arlene, Wayne Koch, Andrew Trotti, and David Sidransky. The New England Journal of Medicine, 2001, 345, 1890 – 1900.
- Herrero, R. Journal of the National Cancer Institute Monographs, 2003, 47 – 51.
- O'Donnell, Rebekah K, Michael Kupferman, S Jack Wei, Sunil Singhal, Randal Weber, Bert O'Malley Jr, Yi Cheng, Mary Putt, Michael Feldman, Barry Ziober, and Ruth J Muschel. Oncogene, 2005, 24, 1244 – 1251.

2

Treatments

- Surgery and Radiation
  - Frequent locoregional reoccurence
- Chemotherapy
  - Cisplatin
  - Irinotecan
  - Paclitaxel
- Wild-type p53 adenoviruses
  - Immunogenic
  - Locoregional reoccurence

- Koch, Wayne M, Joseph A Brennan, Marianna Zahurak, Steven N Goodman, William H Westra, Donna Schwab, George H Yoo, Ding Jen Lee, Arlene A Forastiere, and David Sidransky. Journal of the National Cancer Institute, Nov 6, 1996, 88(21), 1580 – 1586.
- Reed, Andre L, Joseph Califano, Paul Cairns, William H Westra, Richard M Jones, Wayne Koch, Steven Ahrendt, Yolanda Eby, Duane Sewell, Homaira Nawroz, Jiri Bartek, and David Sidranksy. Cancer Research, Aug 15, 1996, 56, 3630 – 3633.
- Pirollo, Kathleen F, et al. Oncogene, 1997, *14*, 1735 – 1746.

3

Problems

- Location of tumors
  - Immunogenic response to treatment
  - Airway obstruction
- Field Cancerization
  - Mucus exhibits mutant p53 proteins
  - Locoregional reoccurrence
- Release of mutant p53 from destroyed cells into blood stream
  - Metastasis

- Perez-Ordonez, B., M. Beauchemin, and R. C. K. Jordan. Journal of Clinical Pathology, 2006, 59, 445 – 453.

4

Possible Solution: Selenium Nanoparticles

- Bacterial growth inhibition
  - Prevent biofilm formation
- Low toxicity in osteoblasts while simultaneously treating osteosarcoma
- Functionalized with PEG → Effective against drug-resistant hepatocellular carcinoma
  - Causes mitochondrial membrane destabilization, resulting in apoptosis
- Temperature-sensitive allotrope shift
  - Particle agglomeration
  - Can be mitigated via dilution and refrigeration

- Wang, Qi and Thomas J Webster. International Journal of Nanomedicine, 2013, 8, 407 – 411.
- Tran, Phong A, Love Sarin, Robert H Hurt and Thomas J Webster. International Journal of Nanomedicine, 2010, 5, 351 – 358.
- Zheng, Shanyuan, Xiaoling Li, Yibo Zhang, Qiang Xie, Yum-Shing Wong, Wenjie Zheng, and Tianfeng Chen. International Journal of Nanomedicine, 2012, 7, 3939 – 3949.
- Dong, Hailong, Aina Quintilla, Marco Cemernjak, Radian Popescu, DagmarGerthsen, Erik Alhswede, and Claus Feldmann. Journal of Colloid and Interface Science, 2014, 415, 103 – 110.

5

Objectives

- Test effect of selenium on viability of head and neck squamous cell carcinoma (HNSCC)
  - Synthesize selenium nanoparticles and characterize
  - Vary dosage concentration and incubation time
  - Compare with effects of selenium on human dermal fibroblasts (HDF)

6

Red Selenium Nanoparticle (rSeNP) Formation

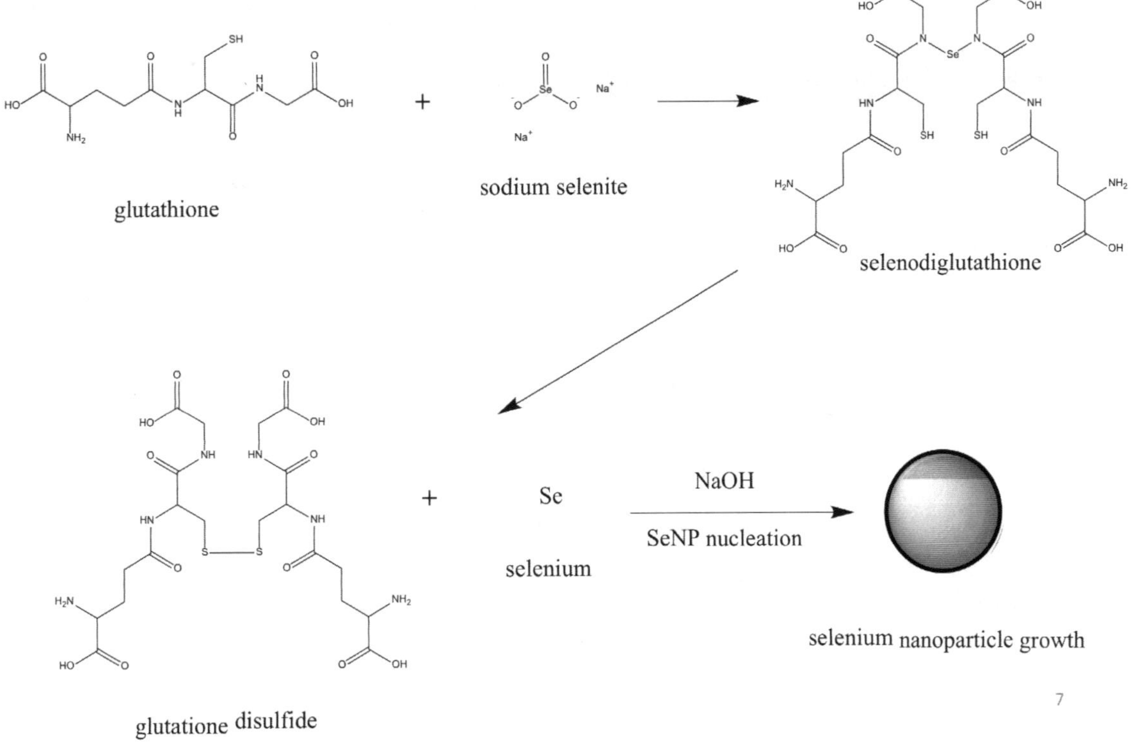

7

rSeNP Characterization - SEM

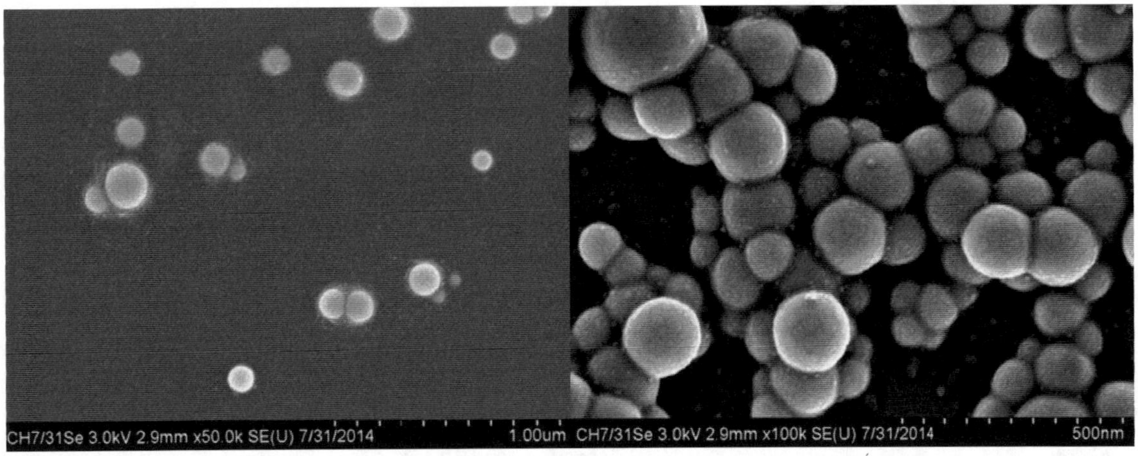

Particle Size = 123.5nm ± 33.6nm

8

rSeNP Characterization – Zeta Potential

- 0.3333 mg/mL rSeNP in PBS
- ζ = -17.54mV ± 5.33mV

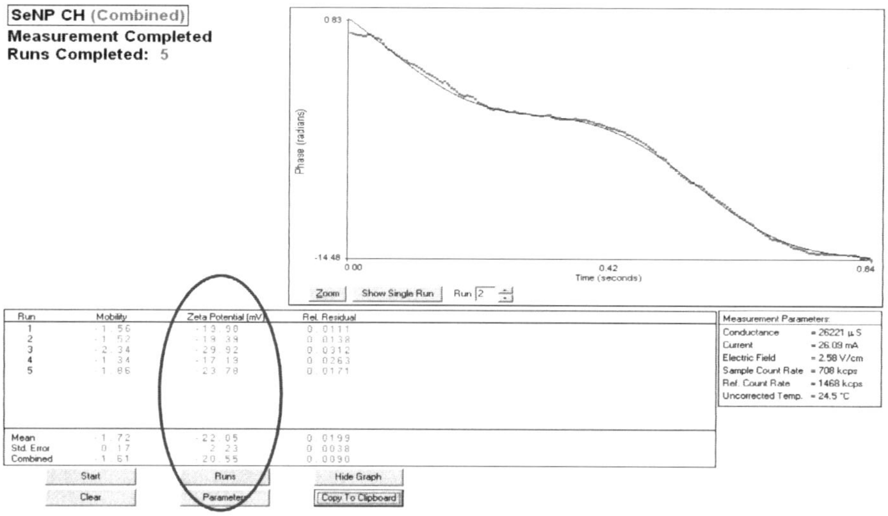

9

rSeNP Characterization - XPS

- Sample found to consist of C, O, Na, and Se

Overall

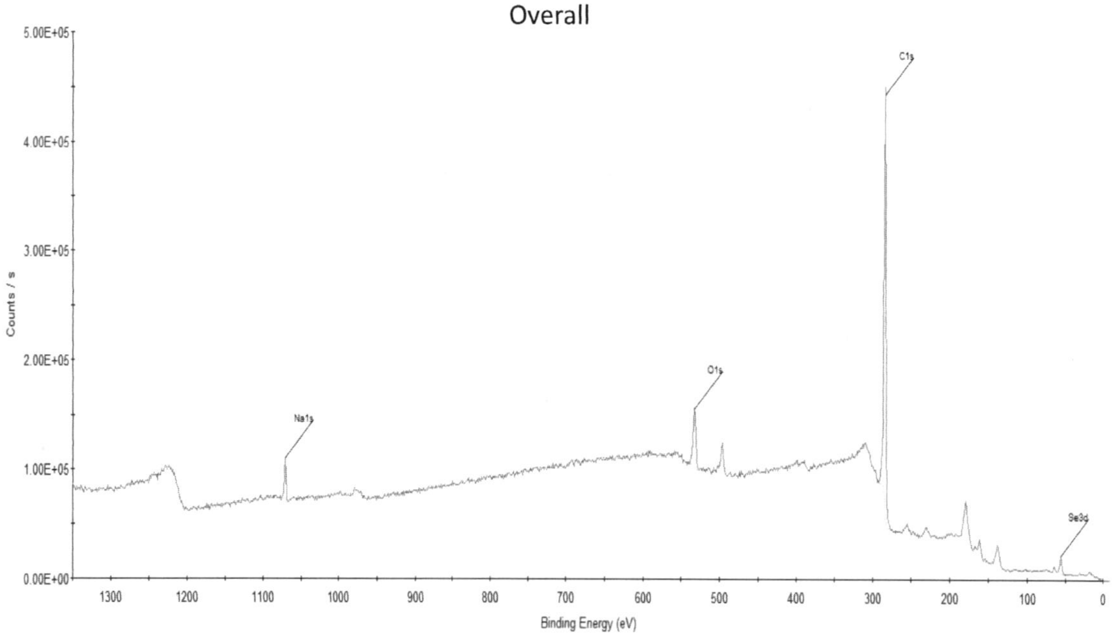

rSeNP Characterization - XPS

- Well-defined selenium peak

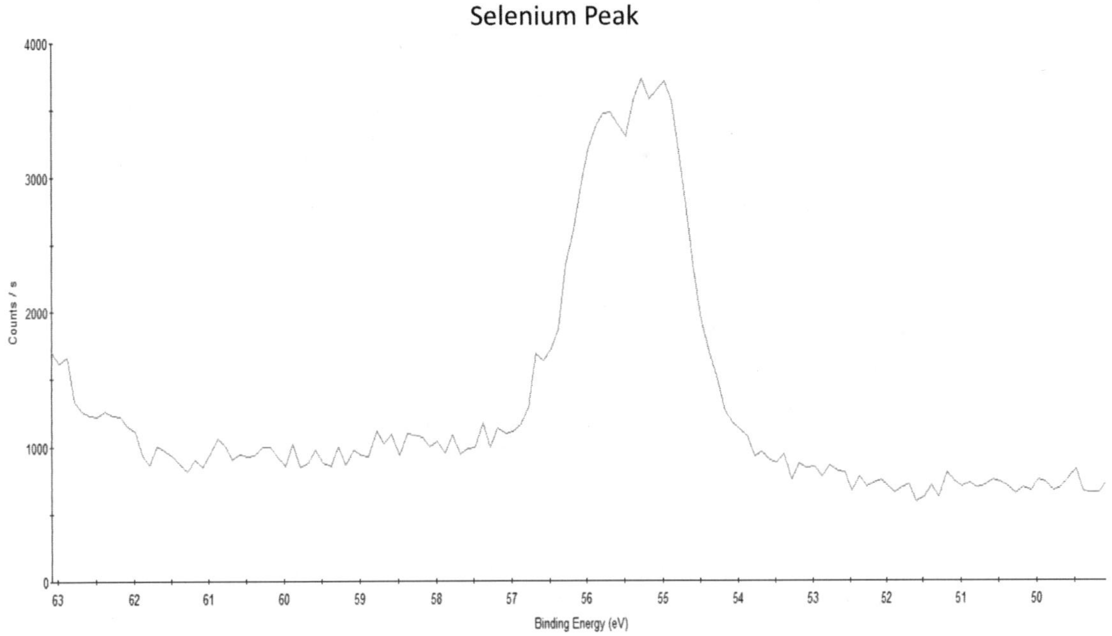

Cell Culture

- Culture cells in 96-well plates at concentrations of 5,000 cells per well.
- Variables
  - Concentration of rSeNP
    - 0.01 – 100 µg/mL media
  - Incubation time
    - 1 – 3 days
  - Perform viability assays with MTS and absorption spectroscopy.
    - Absorbance = 490nm

12

Results
High Toxicity of rSeNP on SCC-9

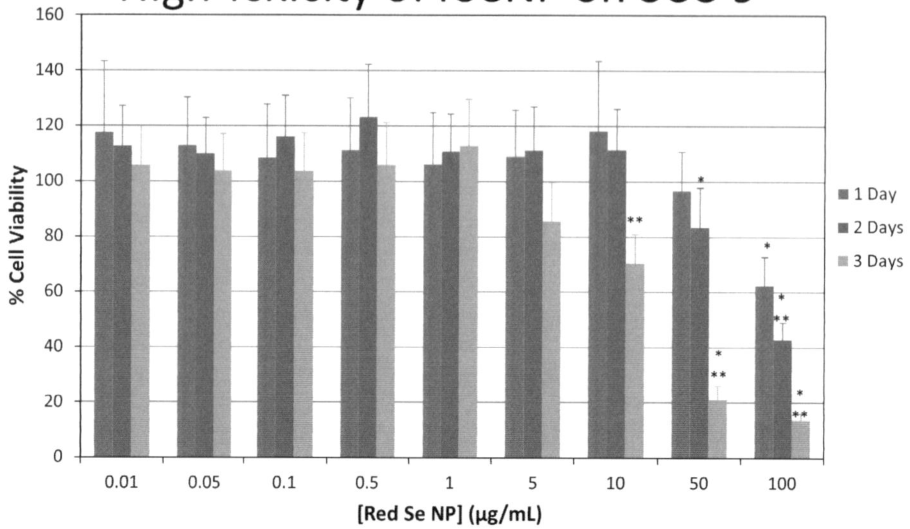

$IC_{50,\ 3Days}$ = 19.2 µg/mL

Data = Mean +/- SEM; N=3; Relative to Untreated Cells
5000 Cells/Well; 120,000 Cells/Data Point
***p<0.001 Compared to Previous Dosage at Constant Incubation Time**
****p<0.001 Compared to Previous Incubation Time at Constant Dosage**

13

Results
Mild Toxicity of rSeNP on HDF

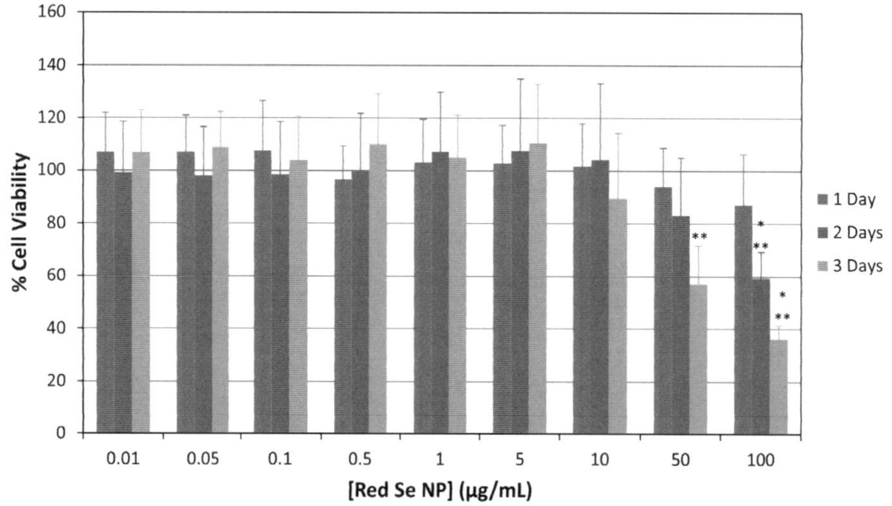

$IC_{50,\ 3days}$ = 59.6 µg/mL

Data = Mean +/- SEM; N=3; Relative to Untreated Cells
5000 Cells/Well; 105,000 Cells/Data Point
***p<0.001 Compared to Previous Dosage at Constant Incubation Time**
****p<0.001 Compared to Previous Incubation Time at Constant Dosage**

14

PEG's Effect on Viability

- PEGylation of rSeNPs shown to help prevent particle agglomeration.
- Increase circulation time.
- Similar viability assays performed using PEG as treatment
  - Concentration of PEG
    - 2.75 – 27500μg/mL media
  - Incubation time
    - 1 – 3 days
  - Perform viability assays with MTS and absorption spectroscopy.
    - Absorbance = 490nm

- Zheng, Shanyuan, Xiaoling Li, Yibo Zhang, Qiang Xie, Yum-Shing Wong, Wenjie Zheng, and Tianfeng Chen. International Journal of Nanomedicine, 2012, 7, 3939 – 3949.
- Knop, Katrin, Richard Hoogenboom, Dagmar Fischer, and Ulrich S. Schubert. Angewandte Chemie International Edition, 2010, 49, 6288 – 6308.

15

Results
Minimal Toxicity of PEG200 on SCC-9

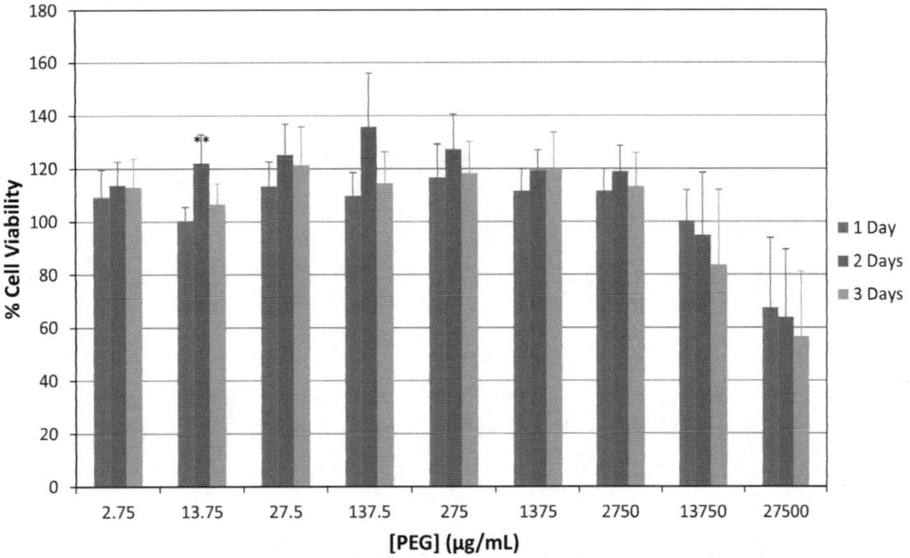

$IC_{50,\ 3Days}$ = 30,100.09 μg/mL

Data = Mean +/- SEM; N=3; Relative to Untreated Cells

5000 Cells/Well; 105,000 Cells/Data Point

****p<0.001 Compared to Previous Incubation Time at Constant Dosage**

16

Results
Negligible Toxicity of PEG200 on HDF

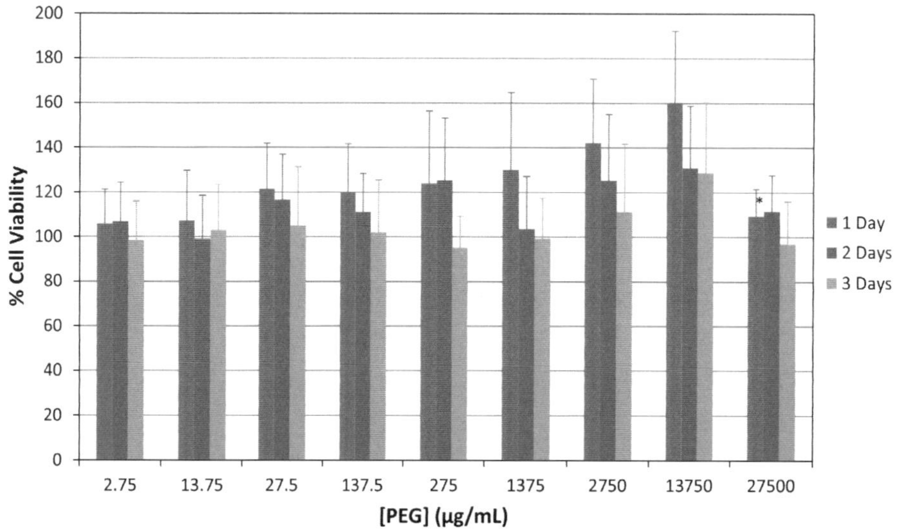

Data = Mean +/- SEM; N=3; Relative to Untreated Cells
5000 Cells/Well; 120,000 Cells/Data Point
***p<0.001 Compared to Previous Dosage at Constant Incubation Time**

17

Discussion and Conclusions

- rSeNPs
  - Significant cytotoxic effect on HNSCC while leaving HDF relatively unharmed.
    - Optimum Range: 20 - 50 μg/mL for 3-day incubation.
- PEG200
  - By itself, slight cytotoxic effect on HNSCC.
  - Negligible effect on HDF.
  - Viable candidate for functionalizing rSeNPs.

18

Future Plans

- Further testing of rSeNPs on the cell lines
  - Cross-reference viability data
- PEG-ylation of rSeNPs
  - Characterization
  - Viability Assays
- Cytotoxicity Pathway Determination
  - Fluorescence Microscopy
  - Laser Scanning Cytometry
- Selective Targeting
  - Target overexpressed proto-oncogenes
    - Epidermal growth factor receptor (EGFR) default target.
  - Further reading to find additional targets.

19

Acknowledgements

- Michelle Stolzoff – Proposed selenium nanoparticle formation mechanism & laboratory procedures
- William Fowle – SEM and TEM
- Harvard University Center for Nanoscale Systems - XPS
- Northeastern University - Funding

Questions?

21

423752 Tripartite Gold Nanoconjugate for Spinal Cord Injury Treatment: Targeting, Recovery, and Biodistribution

Abstract:

Respiratory problems are common and fatal consequences caused by spinal cord injury (SCI) in patients while systemic administrations of drugs that improve respiratory function cause harmful side effects. Gold nanoparticles (AuNPs) show promising potentials in biomedical applications regarding both diagnostic and therapeutic utilities. As a drug delivery carrier, AuNPs enable versatile conjugation chemistry with drugs and proteins and their size, shape, and surface chemistry can be precisely tuned. Despite the progress in nanomedicine our ability to target drugs to regions of disease remains to be very limited. This talk describes a tripartite nanoconjugate comprised of an AuNP drug carrier chemically conjugated to a transporter protein (wheat germ agglutinin coupled to horseradish peroxidase or WGA-HRP) and to SCI drugs such as theophylline. The nanoconjugate was synthesized by first reacting the thiol-protected AuNP with theophylline prodrug to form a biodegradable ester bond linkage. Then WGA-HRP was added to the nanoconjugate by the formation of an amide bond linkage. The final product's chemical composition and nanostructure were characterized by UV-vis spectroscopy, transmission electron microscopy, dynamic light scattering, atomic force microscopy, and thermogravimetric analysis. The nanoconjugate therapeutic efficacy and biodistribution were studied in a validated animal model for hemidiaphragmatic spinal cord paralysis. Our results show that a single administration of the nanoconjugate improved diaphragmatic activity at a much lower dosage than the effective systemic drug dosage and restored the respiratory drive. The effects lasted for 14 days. A biodistribution study using inductively coupled plasma mass spectrometry on the same animal model shows that the nanoconjugate was successful in targeting the respiratory neurons in the medulla. This research demonstrates translational potential for SCI patients with respiratory dysfunction.

Copyright © American Institute of Chemical Engineers. All rights reserved.

WEBSTER NANOMEDICINE LAB
@
NORTHEASTERN UNIVERSITY

Selective Inhibition of Osteosarcoma Cell Proliferation Induced by Self-assembled Peptide Nanospheres Loaded with Curcumin

Kanny (Run) Chang[1], Linlin Sun[1] and Thomas Webster[1,2]

[1]Department of Chemical Engineering, Northeastern University, Boston, MA, USA
[2] Center of Excellence For Advanced Materials Research, King Abdulaziz University, Jeddah, Saudi Arabia

AICHE 2015 Oral Presentation
Nov.9th 2015

1

Osteosarcoma

- Most frequent primary bone cancer caused by abnormal genetic conditions and overexpressed transcription factors
- Exhibits osteoblastic differentiation and produces cancerous osteoid

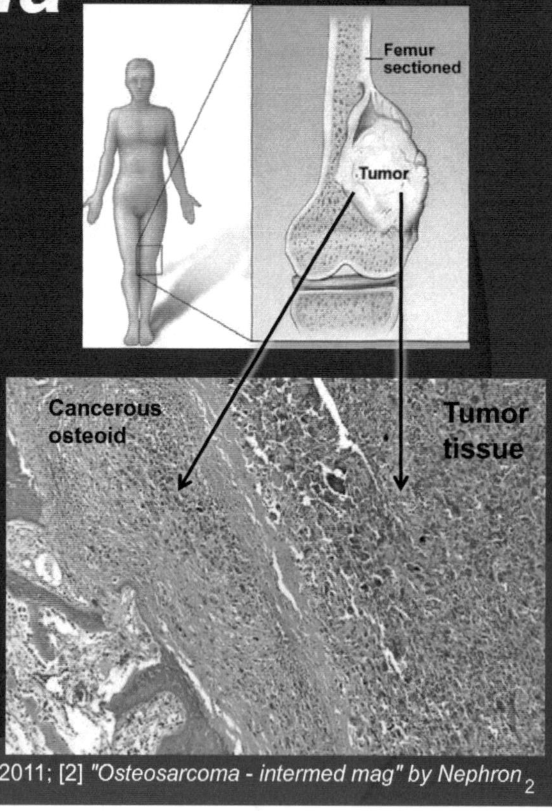

[1] Broadhead. M. L., et al., Sarcoma, Article ID 959248, 2011; [2] *"Osteosarcoma - intermed mag" by Nephron* 2

Chemotherapy Treatment

Integrated strategy consists of:

- Preoperative chemotherapy
- Removal of tumor by amputation or radiotherapy
- Postoperative chemotherapy

Side effects: Kidney and nerve damage (cisplatin), heart damage (doxorubicin), hair loss, and decline of white blood cells and bone marrow cells [1][2]

The approximate survival rate is 65%, but no significant improvements over decades[3]

[1] Arola, O. J., et al. Cancer Research, Vol. 60, pp. 1789-1792, 2000,[2] Broadhead.M. L., et al., Sarcoma, Article ID 959248, 2011,[3] Jaffe, N., et al. Vol. 26, No. 27, pp. 4365-4366, 2008.

3

Curcumin

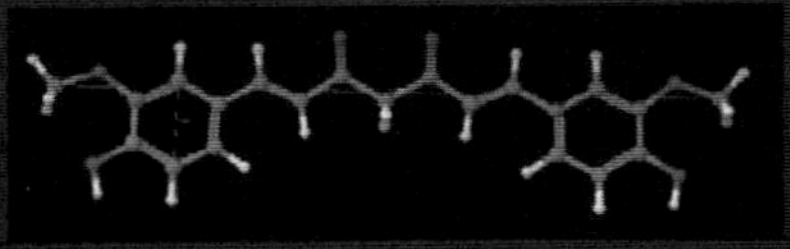

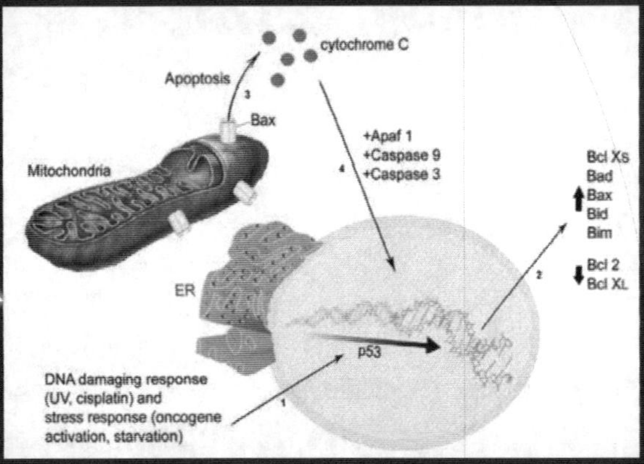

- Orange-yellow colored phenolic compound derived from the plant *Curcuma longa L*
- Shown to inhibit various types of cancer by multiple cellular pathways:

1. *NF-κB transcription factor inhibition*
2. *p53 tumor suppressor activation*

Kumar et al., Critical Review In Therapeutic Drug Carrier Systems, 27(4), pp. 279-312, 2010

4

Carrier design

- Develop drug carrier to deliver curcumin in order to:
  - ✓ *Enhance the solubility of curcumin in aqueous solution*
  - ✓ *Induce higher cellular uptake of curcumin*
  - ✓ *Target osteosarcoma cells*

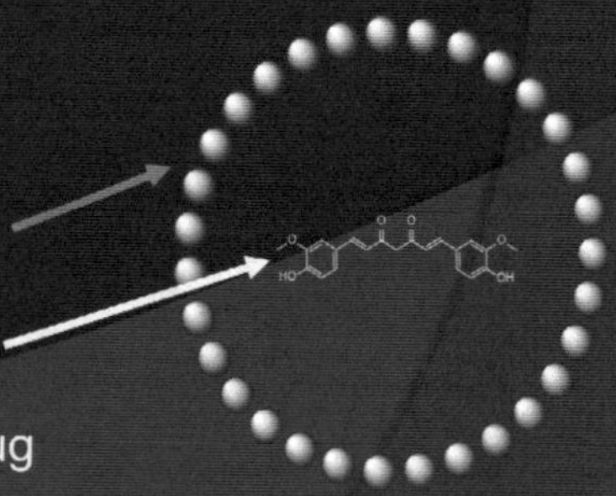

5

Self-assembled Peptide Amphiphiles

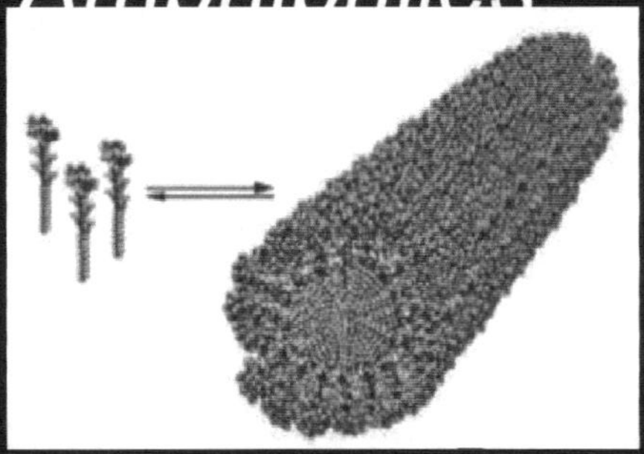

Boekhoven,J., and Stupp. S.I., Adv. Mater. 2014, 26, 1642–1659

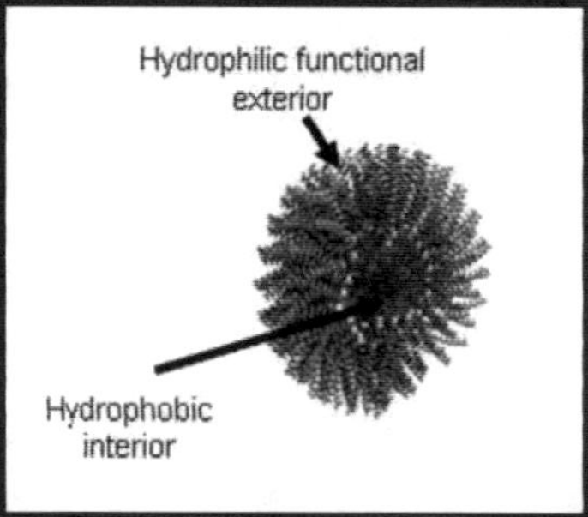

Trent. A., et al., Soft Matter, 2011, 7, 9572

Aliphatic hydrophobic portions :
- *Avoid direct contact with water*
- *Lead to the formation of energetically favorable aggregated cores*

Peptide hydrophilic portions:
- *Offer intermolecular interactions to form supramolecular structures (i.e. H-bond, electrostatic repulsion)*
- *Bioactive motifs can be functionalized for various applications*

7

Amphiphilic Peptide Nanoparticles (APNPs)

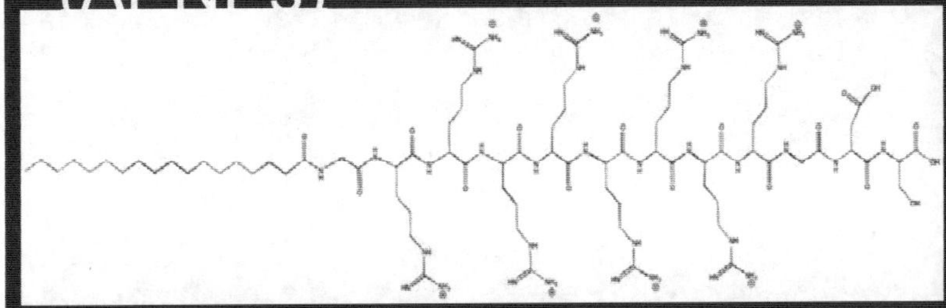

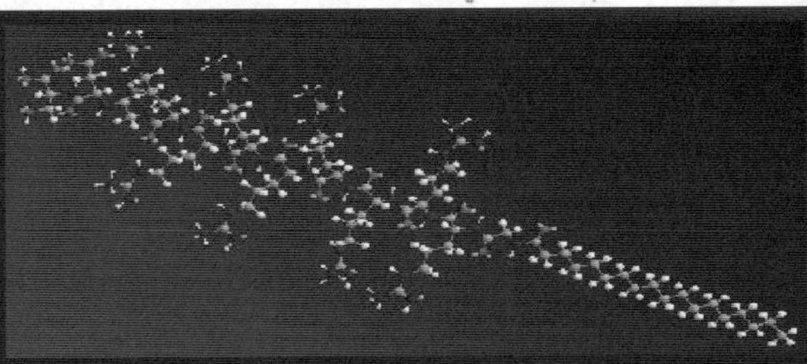

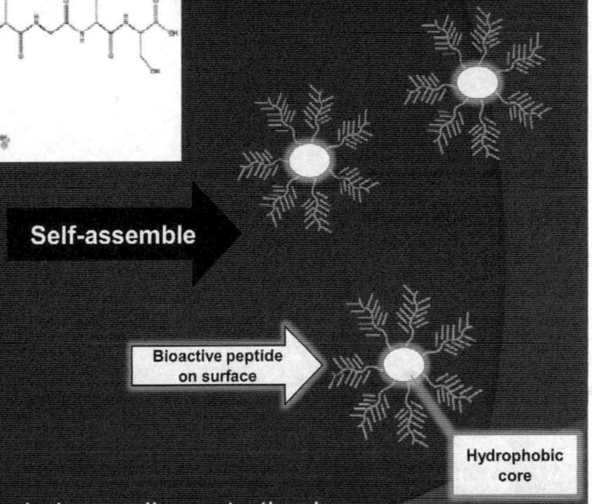

Peptide sequence C17H35-GRRRRRRRRGDS

- The positively charged arginine-rich sequence can induce cell penetration by macropinocytosis, and the RGDS sequence can target the overexpressed integrin receptors on cancer cells
- The fatty acid tail group is the driving force of self-assembly

[1] Chang.R et al. Int J Nanomedicine, Vol. 2015:10, 3351—3365; [2] Chen, et al., Organic & Biomolecular Chemistry, Vol. 8, pp. 3142-3148, 2010.

8

TEM images of APNPs Micelles in Water

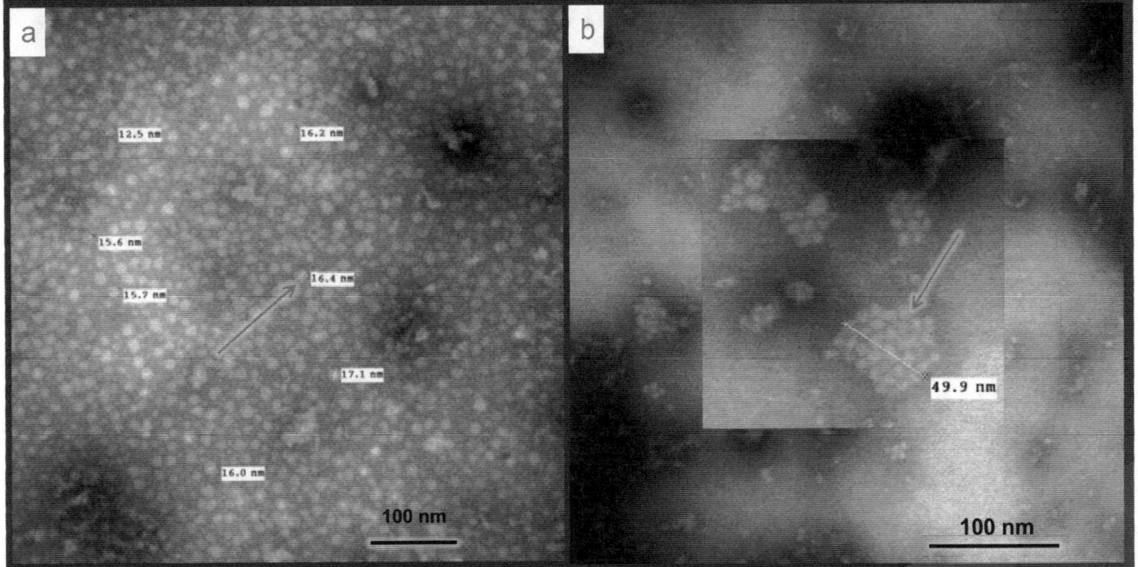

APNPs were formed in globule structure in DI water (a) with sonication and (b) aggregated without sonication. Micelles had diameters from 10-20 nm (average size=15.6 nm)

Samples were stained by 1.5% Phosphotungstic acid (PTA)

9

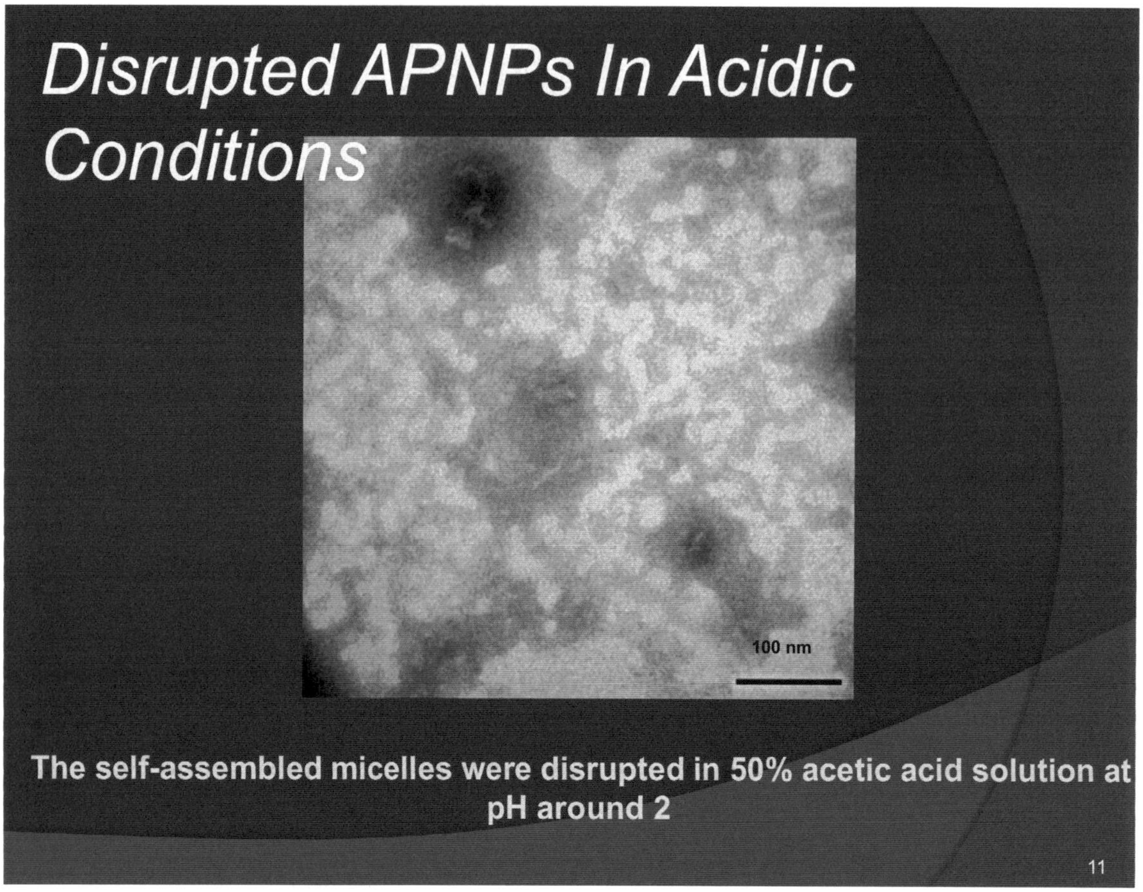
Disrupted APNPs In Acidic Conditions
100 nm
The self-assembled micelles were disrupted in 50% acetic acid solution at pH around 2
11

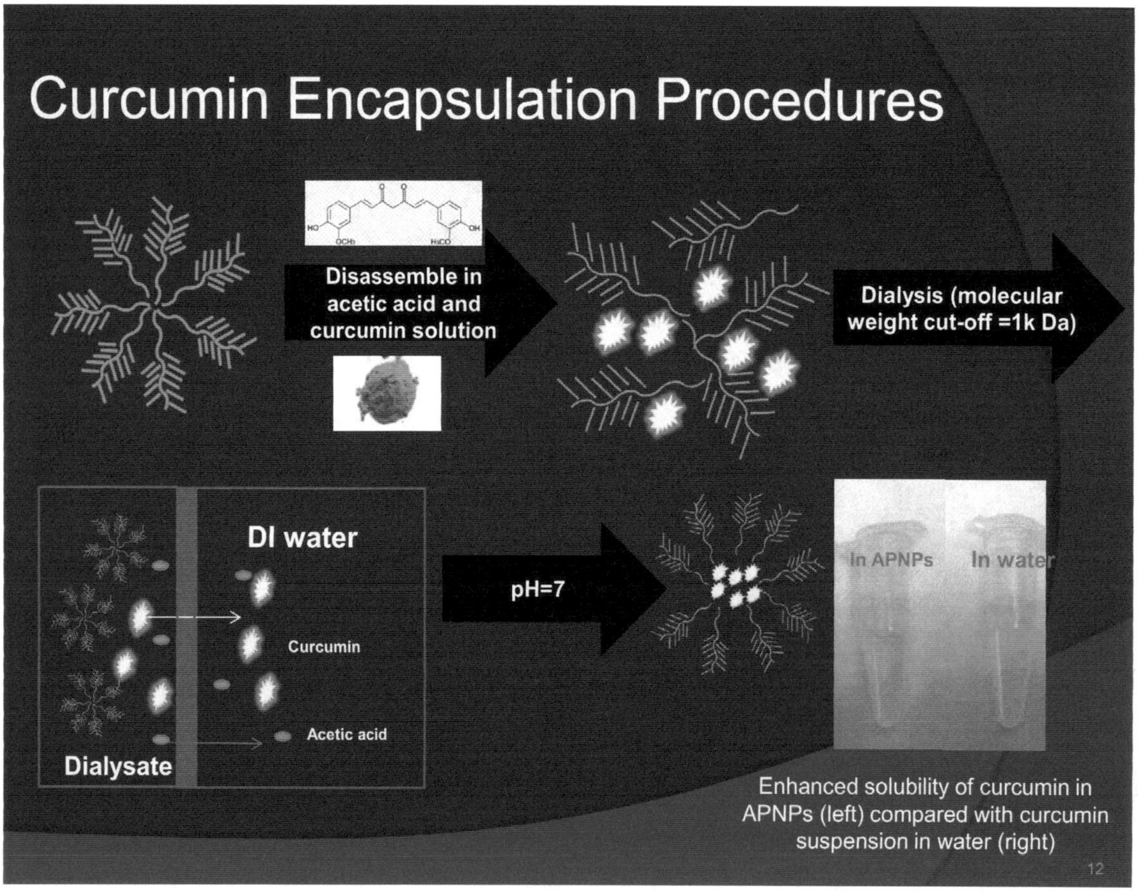
Curcumin Encapsulation Procedures
Disassemble in acetic acid and curcumin solution
Dialysis (molecular weight cut-off =1k Da)
DI water
Curcumin
Acetic acid
Dialysate
pH=7
In APNPs
In wate
Enhanced solubility of curcumin in APNPs (left) compared with curcumin suspension in water (right)
12

Reformation of Micelles in The Post-Dialysis Solution

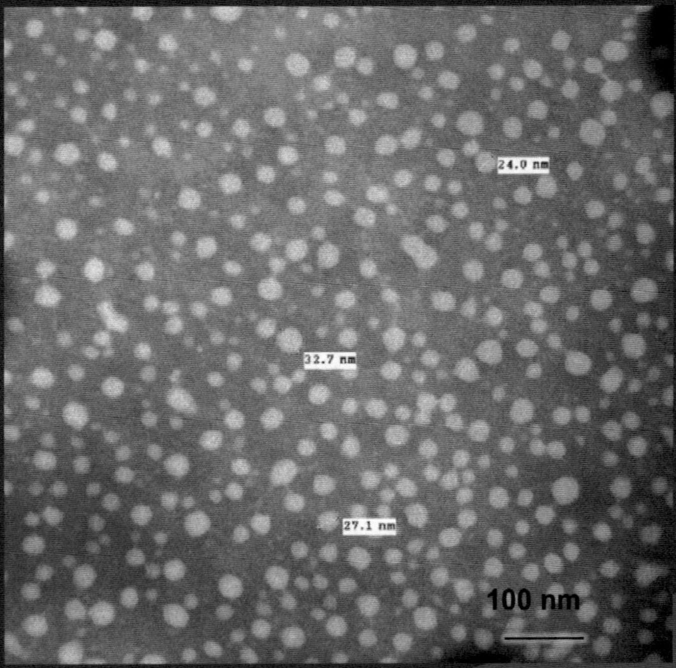

14

FTIR Analysis

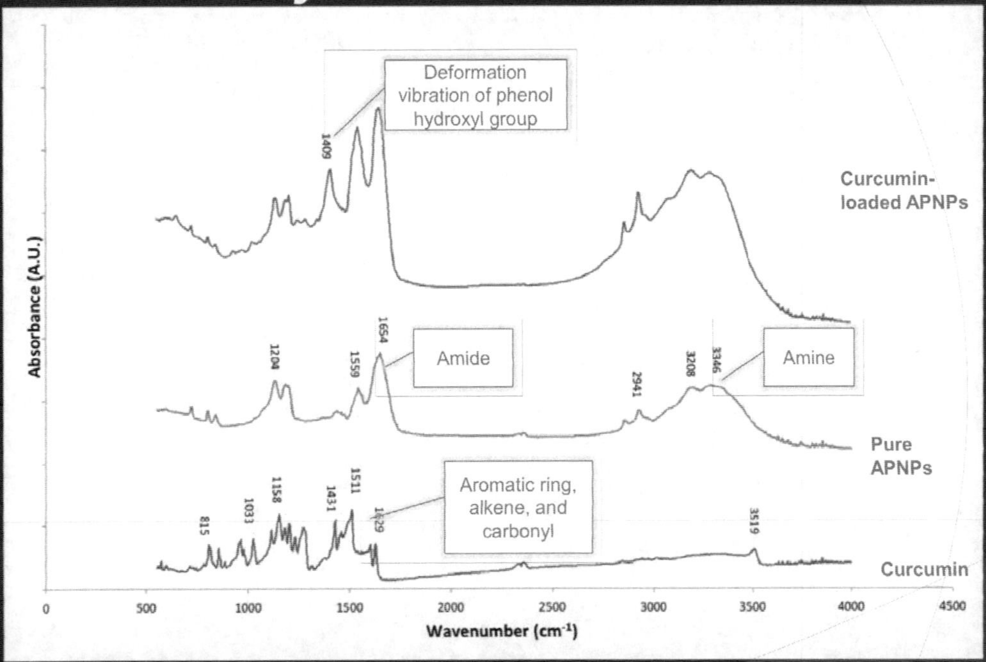

Curcumin may be entrapped in APNPs by physical interaction since no significant band shift was observed

15

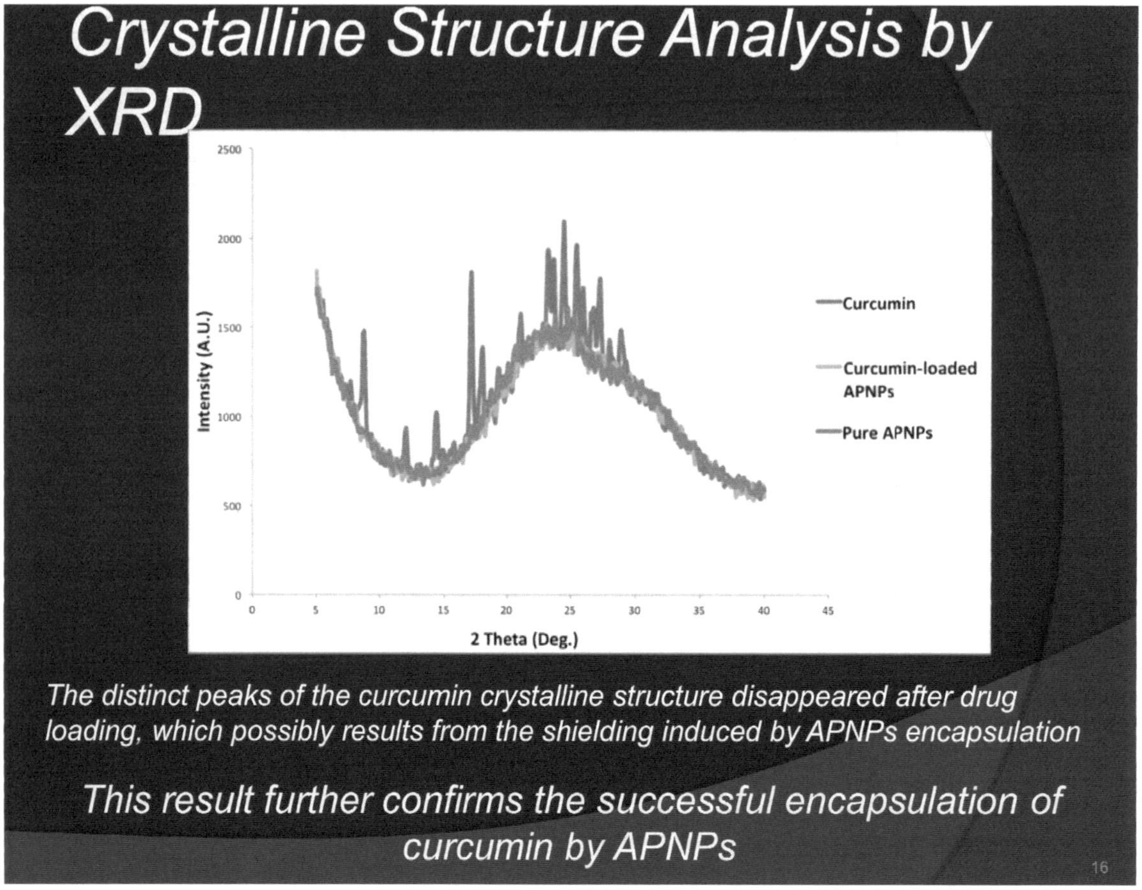
Crystalline Structure Analysis by XRD
2500
2000
1500
1000
500
0
Intensity (A.U.)
0
5
10
15
20
25
30
35
40
45
2 Theta (Deg.)
Curcumin
Curcumin-loaded APNPs
Pure APNPs
The distinct peaks of the curcumin crystalline structure disappeared after drug loading, which possibly results from the shielding induced by APNPs encapsulation
This result further confirms the successful encapsulation of curcumin by APNPs
16

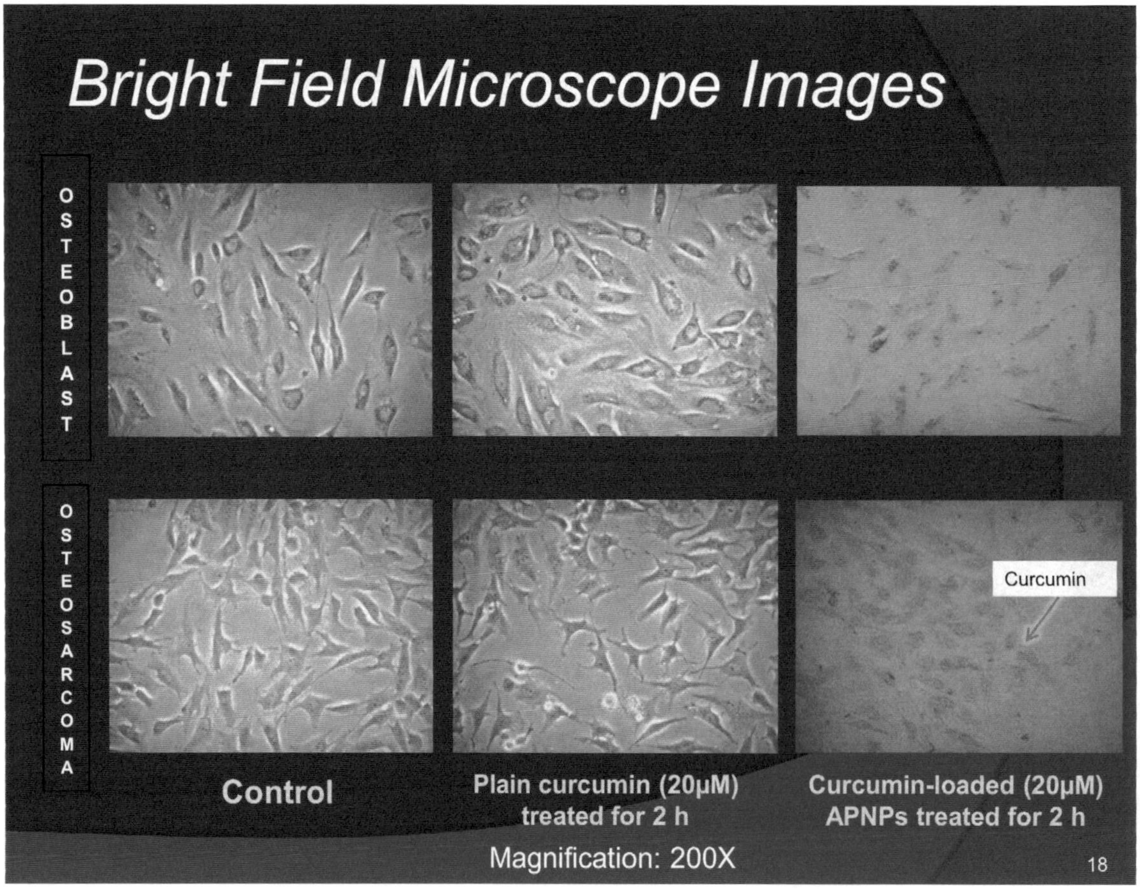
Bright Field Microscope Images
OSTEOBLAST
OSTEOSARCOMA
Curcumin
Control
Plain curcumin (20μM) treated for 2 h
Curcumin-loaded (20μM) APNPs treated for 2 h
Magnification: 200X
18

Confocal Fluorescent Microscope Images

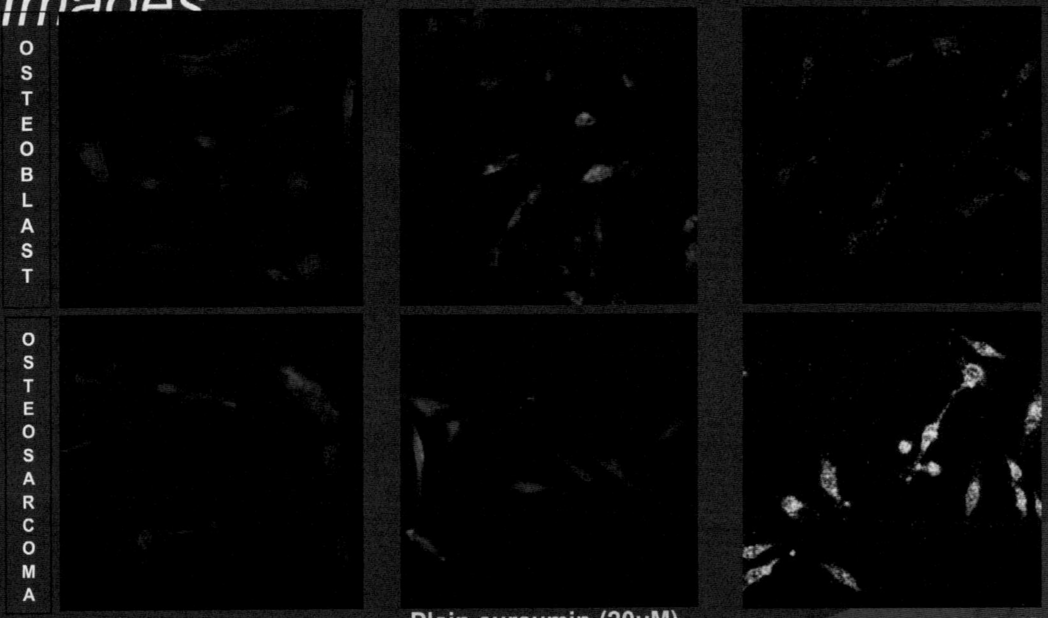

Fluorescent staining: DAPI for nucleus (blue), Rhodamine for F-actin (red), and curcumin was dectected by FITC mode (green). Magnification: 200X.

19

Cytotoxicity of Pure APNPs

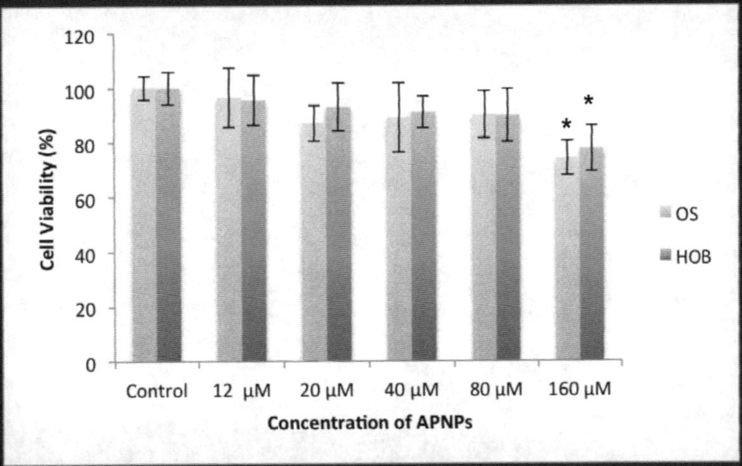

- The cell viability was quantified by MTT assay (a colorimetric assay to quantify cellular metabolic activites)
- Pure APNPs were prepared by the same dialysis as in the drug encapsulation process but without the addition curcumin
- Cells were treated for 24 h with pure APNPs in serum-free medium
- Data are expressed by mean ±S.E.M. and n=3 (5 samples per n). *$P<0.01$ compared with the control samples

20

Cytotoxicity of Curcumin-Loaded APNPs

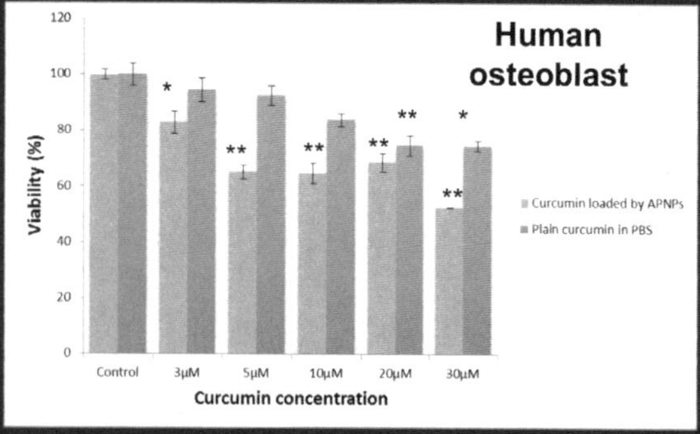

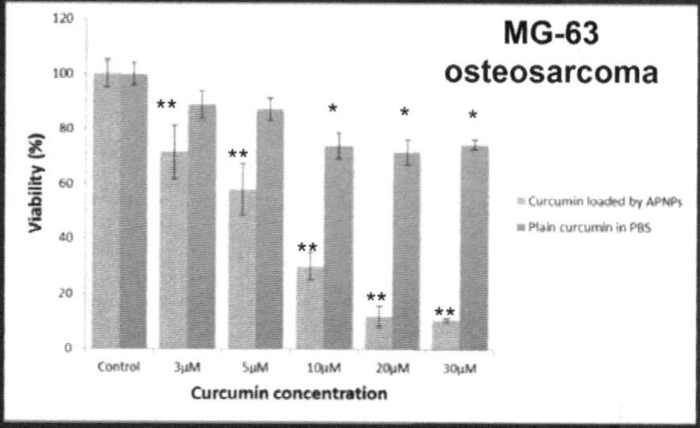

- The cell viability was quantified by MTT assay
- Cells were treated for 24 h with curcumin in PBS and curcumin loaded APNPs in serum-free medium
- Data are expressed by mean ± S.E.M. and n=3 (5 samples per n). *P<0.01, **P<0.005 compared with the control samples

21

Conclusions

- Nanospherical APNPs can be formed with 10-20nm diameters in DI water
- At acidic pH, APNPs disassembled and reassembled at neutral pH
- Hydrophobic curcumin could localize within the hydrophobic cores of APNPs
- The APNPs could target MG-63 osteosarcoma cells more than normal healthy human osteoblast cells

22

Ongoing Work

1. Investigation on mechanisms of the targeting effects of APNPs against osteosarcoma cells

2. Effects of APNPs on other cancer cell lines

1. Improve the stability of APNPs and its ability to encapsulate hydrophobic drugs

23

Acknowledgements

- The authors would like to thank:

1. Lab members in the Nanomedicine Lab
2. Bill Fowle and Rob Eagan
3. Northeastern University for funding

Thank you for your attention
Questions?
25

Skin Permeable Peptide Amphiphiles as an Anti-aging Agent

Gujie Mi[1], Thomas J Webster[1,2]

[1]Department of Chemical Engineering, Northeastern University, Boston, USA.

[2]Center of Excellence for Advanced Materials Research, King Abdulaziz University, Jeddah, Saudi Arabia.

Introduction: One of the most predominant symptoms of skin aging is wrinkle formation, which resulted from both intrinsic aging and environmental damage, such as chronic exposure to UV radiation. Current anti-aging and anti-wrinkle materials often induce a toxic response, which results in inflammation, to increase tissue growth under the skin. Although effective, a much safer and more effective way to alleviate the effect of aging and thereby winkle formation need to be developed. Several biomimetic peptide sequences[1] (KTTKS, GQPR, GHK etc.) have been identified as structural mimics of collagen type I to be effective in stimulating synthesis of key constituents of extracellular matrix by fibroblasts, however, they typically suffer from poor skin penetration due to the existence of stratum corneum. Incorporation of cell penetrating peptides (CPPs) has been an emerging strategy to transport a variety of cargo across cellular membrane in a dose dependent manner.[1] In this study, peptide amphiphiles that consist of both CPPs and biomimetic sequences was designed, synthesized and characterized. The abilities of these peptide amphiphiles to penetrate the skin, to scavenge free radicals, and to promote fibroblast functions (specifically, increasing adhesion, proliferation, collagen synthesis, and decreasing elastase and collagenase synthesis) were determined.

Materials and Methods: Material characterization. Self-assembled amphiphiles were characterized by zeta potential to determine their charge, by dynamic light scattering to examine their size, and by TEM to observe their morphology. Cell viability assay. To determine cell viability, human dermal fibroblasts were seeded at a density of 10,000 cells/cm^2 in 96-well plate and subsequently co-cultured with peptide amphiphiles at various concentrations (12/20/40/80/160 μM) for 24 hours. MTS assays were used to determine cell density after incubation. Briefly, 20 μl of MTS dye solution was added per 100 μl of solution, and the absorbance readings were taken at 490 nm after 4 hours of incubation. Cell density was then determined with correlation to a standard curve. Cell adhesion and proliferation. To determine cellular adhesion/proliferation, keratinocytes and human dermal fibroblasts were seeded at a density of 10,000 cells/cm^2 and cultured respectively in keratinocyte medium supplemented with 10% fetal bovine serum (FBS, Hyclone) and 1% penicillin/streptomycin (P/S, Hyclone), and DMEM supplemented with 10% FBS and 1% P/S. Both types of cells were incubated under standard cell culture conditions (37°C, humidified, 5% CO_2/95% air) for a certain period of time (4h for adhesion; 1,3, and 5 days for proliferation). The MTS assays were used to determine cell density after incubation. Skin penetration/permeation. Skin penetration and permeation was determined using the Static Franz diffusion cell (FD-C) with porcine skin as the membrane between donor and acceptor compartment due to their resemblance to human skin. Peptide amphiphiles were applied in the donor compartment, and PBS was used as the receptor solution. The donor compartment was sealed with aluminum foil and the system was maintained at 37°C in water bath. The skin samples were fixed and sectioned to examine penetration using confocal microscopy. Measurement of total collagens. Sircol soluble collagen assay (Biocolor, UK) was used to quantify total soluble collagens after 7, 14 and 21 days of culturing. All experiments were conducted in triplicate and repeated at least three different times.

Results and Discussion: While peptide amphiphiles showed moderate toxicity at a higher concentration, with LC50 values at around 76.66 μM, they provided improved skin penetration across the stratum corneum when compared to using short peptides alone (Figure 1). Also, it was found that these peptide amphiphiles have the ability to promote cell proliferation as well as total collagen synthesis in the long term.

Figure 1: Density of human dermal fibroblasts after 1 day of culturing when co-cultured with different concentrations of APNP. Cell was seeded at 10,000 cells/cm^2. Values are mean ± SD, n=6. *p<0.005, **p<0.0001 when compared to control group.

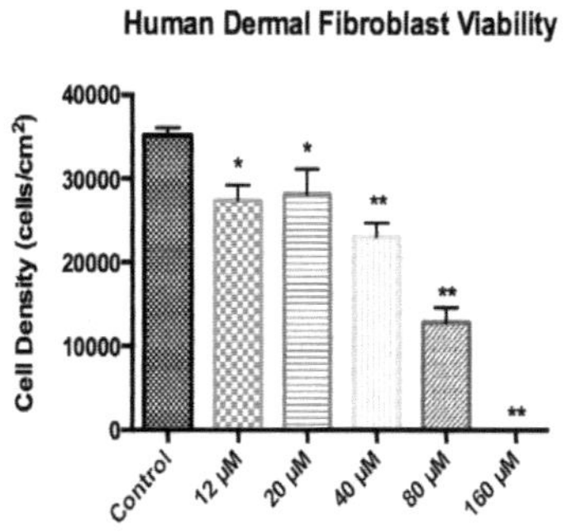

Figure 1: Density of human dermal fibroblasts after 1 day of culturing when co-cultured with different concentrations of APNP. Cell was seeded at 10,000 cells/cm^2. Values are mean ± SD, n=6. *p<0.005, **p<0.0001 when compared to control group.

Conclusions: Through the above experiments, various peptide amphiphiles were identified that can be added to current skin cream formulations to promote penetration of active ingredients, to increase fibroblast growth, as well as to enhance collagen synthesis.

References: [1] Zhao, Xiubo. et al. Molecular self-assembly and applications of desiner peptide amphiphiles. Chem. Soc. Rev., 2010, 39, 3480-3498. [2] Heitz, F. el al. Twenty Years of Cell- Penetrating Peptides: From Molecular Mechanisms to Therapeutics. Br. J. Pharmacol. 2009, 157, 195–206.

426689 Using a Triple Co-Cultured in Vitro Blood- Brain Barrier Modelâ €" to Characterize Magnetic Nanoparticle Permeability

Abstract:

Using a Triple Co-cultured in vitro Blood- Brain Barrier Model to Characterize Magnetic Nanoparticle Permeability

Di Shi[1], Gujie Mi[1], and Thomas J Webster[1*]

[1]Department of Chemical Engineeging, Northeastern University, Boston, USA.

*th.webster@neu.edu

I. Introduction

Among several intriguing nanoparticles, superparamagnetic iron oxide nanoparticles (SPIONs) have been extensively studied because of their ability to be controlled by magnetic fields. For instance, magnetic fields and nanomaterials have been identified to locate and target nucleic acids within the last decade. Moreover, SPIONs have also been discovered as an excellent contrast agent which can potentially increase image contrast and improve MRI sensitivity [1]. Under normal conditions in the body and outside of the brain, the accumulation of magnetic nanoparticles can be removed by several detoxification and antioxidant mechanisms. But, if the SPIONs cross the blood brain barrier and have not been efficiently cleared, the accumulation of those iron oxide nanoparticles may cause neurodegeneration and be harmful to normal brain function in the long term [2].

As there are concerns about both the delivery of drugs across the blood-brain barrier and the accumulation of SPIONs in the body, this study established a triple co-cultured *in vitro* blood-brain barrier model to test the ability of a new type of magnetic material (magnetic nanoliposome) to pass through the blood-brain barrier. We tested several variations of magnetic nanopliposom in an effort to both increase blood-brain barrier passage (for neural drug delivery applications) and decrease blood-brain barrier passage (to minimize toxicity).

II. Materials and Methods

<>A. Material Characterization

The collagen coated nanoliposome and PEG coated nanoliposom were prepared and encapsuled with iron oxide [3]. The nanoparticles will then be characterized by zeta potential to determine their charge and dynamic light scattering for hydrodynamic diameter. SEM was used to characterize their surface morphology and TEM was used to assess the iron oxide inner core diameter.

<>B. Experimental Samples Tested Through the Blood-Brain Barrier Model

For the in vitro blood- brain barrier model, pericytes were cultured and seeded on the bottom side of the collagen-coated Transwell¨ inserts with 104 cells/cm2 density and astrocytes with same density were seeded on the bottom of the 24-well plates separately. After 12 hours of adhesion, murine brain endothelial cells were seeded onto the upper side of the inserts with 105 cells/cm2 density and the inserts would then be placed in the 24-well plates containing astrocytes. The model would be evaluated and confirmed using TEER and FITC-Dextran transport. Then, the model was used to test the permeability of the various nanoparticles after confirmation. The in vitro model would be exposed to nanoparticles for 2 hours. After 2 hours, a 100 μL solution was taken from each well and an iron assay kit was used to determine the iron concentration that passed through the model. Each experiment was conducted in triplicate and repeated at least three times.

III. Results and Discussion

<>A. Material Characterization

Dynamic light scattering revealed that the hydrodynamic diameters of the samples ranged from 20nm to 200nm. TEM images show that the iron core was about 5-10nm for all of the samples.

<>B. Experimental Samples Tested Through the Blood-Brain Barrier Model

The permeability of FITC-Dextran across the model confirmed that the model was successfully established and the TEER value of this triple co-cultured model was around 350 Ohms/cm^2. Previous results of samples showed that the highest permeability was obtained from collagen coated nanoparticles. This result suggest that nanoliposomes coated with collagen had better permeability across the blood-brain barrier than nanoliposomes coated with PEG. Through such experiments, magnetic nanomaterials (such as megnetic nanoliposome) suitable for MRI use which are less permeable to the blood brain barrier to avoid neural tissue toxicity and magnetic nanoliposomes suitable for brain drug delivery since they were more permeable to the blood-brain barrier were created [4,5].

IV. Conclusions

An *in vitro* model of blood-brain barrier was established using triple co-culture method. The model was confirmed by comparing the permeability trend of FITC-dextran in serum-free medium and the TEER value with previous research. The results suggest a possibility to manipulate magnetic nanoparticle penetration across the blood–brain barrier by controlling bioactive coatings. Such data lay the foundation for the modification of ferrofluids to be either coated with collagen to pass through blood-brain barrier, or to be coated with glycine and glutamic acid to avoid penetration.

Acknowledgment

The authors would like to thank Northeastern University for funding this research.

<>References

[1] Christian P, et al. Magnetically enhanced nucleic acid delivery. Ten years of magnetofection—Progress and prospects, Advanced Drug Delivery Reviews. 2011; 63: 1300-1331.

[2] Richard G, et al. BOLD functional MRI in disease and pharmacological studies: room for improvement?, Magnetic Resonance Imaging, 2007; 25: 978-988.

[3] Krishnamoorthy G, et al. Collagen Coated Nanoliposome as a Targeted and Controlled Drug Delivery System. AIP Conf. Proc, 2010; 1276, 163.

[4] Di S, et al. Controlling ferrofluid permeability across the blood–brain barrier model. Nanotechnology, 2014; 25 (7), 075101.

[5] Bennett J, et al. Blood–brain barrier disruption and enhanced vascular permeability in the multiple sclerosis model EAE. J Neuroimmunol. 2010; 229:180–191.

Copyright © American Institute of Chemical Engineers. All rights reserved.

434573 Size and Shape Characterization of Hydrated and Desiccated Exosomes

Abstract:

Cells actively secrete nano-vesicles called exosomes via an endocytic pathway. Exosomes are found in the extracellular space and all body fluids, including blood, urine, and saliva. The molecular content of exosomes is derived from the cells that release them. The frequency and amount released depends on the cell type, such as normal or cancerous, and environment, such as the level of dissolved oxygen and pH. They carry small non-coding RNAs, though the number of microRNA (miRNA) molecules enveloped inside each exosome may vary. Some reports suggest that the majority of miRNAs in the blood are contained within exosomes. By fusing with recipient cells and releasing their RNA and other cargo, exosomes are thought to play a role in short and long-range intercellular signaling.

Exosomes can be differentiated from other circulating vesicles by the markers of the endosomal pathway and their small size that distinguishes them from other microvesicles. Though an important differentiator, their reported size and shape vary substantially, which likely reflects the difference in cells that secreted them, utilized isolation techniques and the methods used in their characterization. We analyzed the influence of the characterization method on the measured size and shape of hydrated and desiccated exosomes. It was found that hydrated exosomes are close to spherical with a hydrodynamic size that is significantly larger than the geometric size. In addition, surface fixation prior to analysis in liquid that is required for techniques such as atomic force microscopy significantly impacts the size and shape of exosomes. For desiccated exosomes, it was found that the shape and sizing are influenced by the manner in which drying occurred. Isotropic desiccation in aerosol preserves the near spherical shape of the exosomes, while drying on a surface likely distorts their shapes and influences sizing results.

Copyright © American Institute of Chemical Engineers. All rights reserved.

415045 Novel Light-Activated Nano-Therapeutics for Selective Cell Phenotypes

Abstract:

Current design of therapeutics (drugs and nanoparticle-based therapies) relies heavily on a spatially selective drug-delivery system (colloids and vesicles coated with desired antibodies or proteins) due to non-specificity of the drugs or metal nanoparticles. Typically, these non-specific drugs once spatially delivered kill the targeted cells (e.g. tumor or pathogen) as well as healthy cells in close proximity (including metal nanopartciles used for cancer therapy). Further complexities in non-specific binding and incorrect delivery can cause undesirable side-effects. Here we present a novel nanoparticle-based therapy where the (non-metallic) nanoparticles specifically target selective cell phenotypes, based on their biochemical potentials. Using a specific light induced targeting mechanism, these light-activated reactive species (LARS) have the capability to target and kill specific-pathogens on illumination with visible light, without affecting the growth of other cells. While the growth of selected cells and the targeted pathogens is unaffected in dark, we selectively designed different LARS species which can either kill or proliferate the growth of selected target cells. This is demonstrated when targeting *Escherichia coli* cells, HEK293 mammalian cells, as well as, a co-culture of *E. coli* and HEK293 cells. Furthermore, within the same strain of *E.coli*, we find that while lab strain DH5α is susceptible to LARS, clinically isolated multi-drug resistant clinical strain of *E. coli* is not; likely due to presence of a non-specific antibiotic resistance mechanism. The ability to tune the biochemical potential of nanoparticles allows for selective cell phenotypes over a range of organisms, and has potential applications in antimicrobials and targeted organism-specific therapeutics.

Copyright © American Institute of Chemical Engineers. All rights reserved.

437507 Challenge and Promise of Extracellular Nanovesicles in Early Detection and Treatment of Cancer and Other Systemic Diseases

Abstract:

Exosomes and other nano-size extracellular vesicles (EVs) are stable membrane-bound nanoparticles secreted into the circulation by diverse cell types, including epithelial cells, mesenchymal cells, lymphocytes, and tumor cells. They are 20-120 nm in diameter and contain membrane and cargo, which includes nucleic acids and proteins, originating from the parent cell. They are found in astronomical numbers in extracellular spaces and body fluids, including blood, urine, saliva, breast milk, semen, lymphatic, and spinal fluids. In blood alone, there are over 10^{16} circulating exosomes, the physiological functions of which are not completely understood. It has already been shown that exosomes are important in modulating immune response and are thought to play a role in short and long-range intercellular signaling by fusing with recipient cells and releasing their cargo. Their population is heterogeneous, reflecting the diversity of cells that secrete them which, in patients, include abnormal (e.g. malignant) cells. The stability of these vesicles allows their isolation from body fluids, characterization by molecular content and biophysical properties, which together can potentially be used for medical diagnostics. However, the fundamental difficulty in using molecular content of exosomes in early and routine diagnostics is the large background of nanovesicles originating from normal cells found in the circulation that interferes with the analysis even when the most sensitive techniques (such as digital droplet PCR) are used. Our laboratory has focused on identifying biophysical properties of cancer exosomes that most distinguish them from other nanovesicles in the blood and use these properties to fractionate the sample and enrich in tumor exosomes. In this approach, we aim to decrease the influence of the background in order to improve the accuracy of downstream molecular measurements, ultimately improving the sensitivity to allow for an early detection of cancer and other systemic diseases.

Copyright © American Institute of Chemical Engineers. All rights reserved.

437537 Contributions Chemical Engineers Can Make to Therapeutic Delivery to the Lungs

Abstract:

As Chemical Engineers we sometimes wonder what we can contribute to medical science. I will give examples from our work, and suggest that the first step is collaborating with medical researchers who understand the physiology and disease pathology that needs to be addressed. With that starting point our understanding of fluid mechanics, material science, and self-assembly enables us to make important contributions which the medical community is not equipped to provide. We will give two examples of nanoparticle delivery to the lungs that address problems in TB and non-small cell lung cancer. The first example is the creation of inhalable aerosol particles comprising therapeutic nanoparticles and a matrix phase to create low-density particles that enable delivery to the deep lung. The second example is the creation of gel microparticles that deliver nanoparticles through venous filtration. The gel microparticles are created by emulsion processing and polymer gelation so that the IV injected particles are trapped in the vasculature in the lungs where they deliver anti-cancer drugs over extended periods.

Copyright © American Institute of Chemical Engineers. All rights reserved.

430000 Separation of Bacteria from Blood Components

Abstract:

Blood bacteremia, or a bacteria infection in the blood stream, is of major concern in terms of mortality and morbidity, particularly when antibiotic resistant bacteria are involved. Persons presenting bacteremia symptoms are often given standard doses of general antibiotics and a blood sample is taken to culture the bacteria for identification. The problem is that culturing and identifying the pathogen in order to administer the correct antibiotic response usually requires 24 to 48 hours, during which time the patient could become severely sick if the particular pathogen is resistant to the general antibiotic already administered.

Our overall goal is to develop methods that are much faster, even as fast as one hour, that can identify antibiotic resistant bacteria from a person's blood. This implies that there is no time to culture the bacteria in the blood to obtain sufficient numbers of bacteria to identify resistance by current conventional means. We are developing methods to quickly identify the bacteria in the blood using molecular mechanisms such as direct DNA or RNA identification without PCR amplification of certain gene sequences and without culture amplification of the bacteria. The first step in this process is to separate the bacteria from the blood components that would interfere with the subsequent assays. The means separating the bacteria from the red blood cells (RBCs), white blood cells (WBCs) and platelets (PLTs). The target time is 10 minutes or less for separation, after which the recovered bacteria will be lysed and their DNA analyzed for identification within 50 minutes.

This submission presents our successes and failures in using porous hollow fibers, flat membrane filters and centrifugal processes for separation.

While the use of porous hollow fibers with 3-5 um pores is good in theory, the application is difficult because hollow fibers with uniform cylindrical pores are not commercially available. Porous hollow fibers with elliptical pores of 5 um size are available, but these trap bacteria (not pass bacteria) because of the pore shape. Pores with tortuous pathways also trap bacteria that would otherwise pass through a cylindrical pore.

Flat membranes with uniform pores are commercially available, and have been used in this work with bovine blood. However, the bacteria recovery is low, and the area needed to prevent filter cake build-up is so large that a small disposable device that could be used in a hospital becomes impractical in terms of size and expense.

Centrifugal separation can easily separate RBCs (~8 um) and WBCs (~15 um) from platelets (~3 um) and bacteria (0.5 to 2 um) because of the large size of the cells. The dynamic sedimentation velocity goes as the diameter squared, and red cells are relatively dense because of their hemoglobin content. In these preliminary studies we have used bovine blood spiked with E. coli. We have had success in centrifugal separation with high bacterial recovery as only a small fraction of bacteria gets trapped in packed cells. We have developed and will present some unique centrifugation procedures that increase the efficiency or separation of cells from bacteria.

Copyright © American Institute of Chemical Engineers. All rights reserved.

437510 Engineering of Articular Cartilage: Challenges and Prospects

Abstract:

Engineering of articular cartilage: Challenges and prospects

Nehal Abu-Lail, Chrystal R. Quisenberry, Arshan Nazempour, and Bernard Van Wie

Voiland School of Chemical Engineering & Bioengineering, Washington State University, Pullman, WA 99164-6515.

Introduction: Articular cartilage (AC) is the smooth white connective tissue of diarthrodial joints. It acts as a lubricant to the moving joins as well as a distributor of loads applied on them. However, AC tissues are subjected to extreme mechanical loads, harsh chemical conditions and worse of all, they are devoid of blood and nerve vessels. Mainly due to the latter, AC tissues have limited regenerative capabilities when injured or damaged. AC failure results in osteoarthritis (OA) or can cause musculoskeletal morbidity. Currently, no successful long-term treatments of OA are present. Patients with OA usually have to replace joints. In the US alone, a joint is replaced ~ every 90 seconds and it is even worse in Europe. With no treatments, AC presents an ideal candidate for tissue engineering. Tissue engineering is an interdisciplinary field in which biological concepts are combined with engineering tools and quantitative methods to model environments capable of producing functional tissues that can be used to replace native diseased tissues. Here, we are utilizing a unique centrifugal bioreactor, an array of mechanical and chemical cues and a variety of cell types to engineer AC *in vitro.* We hypothesize that biomechanical and chemical stimulation will be sensed by the cell *via* mechanotransduction through β1-integrins and N-cadherins and this will prompt cells to secrete extracellular matrix. We further hypothesize that this stimulation will improve the Young's modulus of cartilage tissues by increasing it to near native cartilage properties.

Materials and Methods

Chondrocytes, the primary cell type in AC, remodel the tissue's extracellular matrix (ECM) in response to mechanochemical cues from their environment. Unfortunately, during growth, they dedifferentiate. With that limit, human adipose-derived stem cells (hASCs) or bone marrow stem cells can be used as a cell source. Cells were seeded in our novel centrifugal bioreactor (CBR) in a variety of ways. To mimic native AC environment, oscillating hydrostatic pressure and transforming growth factor-β3 (TGF-β3) were also applied. These cues lead to differentiation of stem cells into chondrocytes to form a cartilage tissue with desirable properties. Cells were generally grown for 21 days in the reactor as well as in static controls of micromass and pellet. The functionality of the engineered AC tissues was tested via an array of techniques. First, atomic force microscopy (AFM) was used to measure the mechanical properties of the cell and cartilage tissue as well as to map the location and density of β1-integrins on the cell surface as a measure of mechanotransduction. mRNA sequencing was used to probe for 10 markers of chondrogenesis. Finally, biochemical analyses were carried to quantify DNA and glycosaminoglycans (GAG) contents of the engineered cartilage.

Results and Discussion: CBR tissues grown with only oscillating hydrostatic pressure had Young's moduli 44-fold higher than the micromass or pellet static controls. With the addition of TGF-β3 to the CBR tissues, a 1.8-fold increase was observed, bringing the Young's modulus nearer to that of native cartilage. It was found that the average adhesion force for a single antibody-β1-integrin interaction was 80 pN, independent of treatment group. The micromass and pellet cultures had two to five-fold larger β1-integrin counts compared to the cultures engineered in the CBR. Because β1-integrins facilitate cell-ECM attachment, a deficient ECM may require more β1-integrins to attach to the cell. Analysis of N-cadherins on the surfaces of cells in response to stresses is ongoing.

Conclusions: Cartilage tissues grown from hASCs had higher Young's moduli, nearing that of native cartilage, when oscillating hydrostatic pressure and TGF-β3 were applied synergistically. The surface expression of β1-integrin was dependent on the tissue culture method. In the future, adding a scaffold with gradients in its properties to the CBR may bring us a step close to engineering AC with properties close to those of native cartilage.

Acknowledgements: This work was supported by an NSF EAGER grant 1212573, the NIH Protein Biotechnology Training Program 24280305, a NASA Space Grant, a WSU DRADS fellowship, and a Harold P. Curtis Scholarship.

Copyright © American Institute of Chemical Engineers. All rights reserved.

437505 15 Years of Chemical Engineering Advances in Medicine

Abstract:

There is an acute shortage of organs due to disease, trauma, congenital defects, and most importantly, age related maladies. The synthetic materials used in tissue engineering applications today are typically composed of millimeter or micron sized particles and/or fiber dimensions. Although human cells are on the micron scale, their individual components, e.g. proteins, are composed of nanometer features. By modifying only the nanofeatures on material surfaces without changing surface chemistry, it is possible to increase tissue growth of any human tissue by controlling the endogenous adsorption of adhesive proteins onto the material surface. In addition, our group has shown that these same nanofeatures and nano-modifications can reduce bacterial growth without using antibiotics, which may further accelerate the growth of antibiotic resistant microbes. Inflammation can also be decreased through the use of nanomaterials. Finally, nanomedicine has been shown to stimulate the growth and differentiation of stem cells, which may someday be used to treat incurable disorders, such as neural damage. This strategy also accelerates FDA approval and commercialization efforts since new chemistries are not proposed, rather chemistries already approved by the FDA with altered nanoscale features. This invited talk will highlight some of the advancements and emphasize current nanomaterials approved by the FDA for human implantation. It will also provide a perspective on the role chemical engineers have played in the field of nanomedicine, in terms of pioneering the field, contributing state-of-the-art knowledge, commercialization, and education.

Copyright © American Institute of Chemical Engineers. All rights reserved.

437364 Detection of Mycobacterium Tuberculosis Volatile Biomarkers Using a Titanium Dioxide Nanotube Based Sensing Platform

Abstract:

This research presents a metal functionalized titanium dioxide (TiO_2) nanotube based sensor developed for detection of volatile biomarkers (methyl *p*-anisate and methyl nicotinate) associated with tuberculosis (TB). In this work self-aligning TiO_2 nanotubes were fabricated, and functionalized with cobalt for specific binding to the volatile biomarkers of interest. Mimics of the volatile biomarkers were dissolved in water at variable concentrations, and nitrogen gas was flowed through the biomarker solutions and delivered to the sensor. A potentiostatic scan was used to measure current across the device under constant potentials ranging from -1.0 volts to 1-volts methyl nicotinate and methyl *p*-anisate. The sensor response showed a 3×10^5 change in current for methyl nicotinate, and 7.61×10^6 change in current for methyl *p*-anisate when compared to baseline. The sensor also showed little response to five other VOCs (acetone, ethanol, methanol, benzene, and phenol) which suggest the sensor is functionalized to be specific for the target VOCs associated with TB. Results showed detection of the volatile biomarkers yield orders of magnitude change in current which is detected easily. The sensor is simple to operate, responds in minutes, and has potential applications in non-invasive diagnosis of TB and other diseases that have distinct volatile biomarkers.

Copyright © American Institute of Chemical Engineers. All rights reserved.

437367 Synthetic Fascia

Abstract:

Failure of a soft connective tissue called fascia is at the heart of a number of debilitating medical maladies. For example, over a half million surgeries in the US are performed to correct pelvic organ prolapse and urinary incontinence each year. However, currently available synthetic alternatives available in sufficient quantities do not match the materials properties of adjacent tissue leading to its erosion and may slip after insertion leading to device failure. This talk will highlight our efforts to design synthetic fascia that matches the material properties of adjacent tissue and uses nanomaterials to minimize device slip.

Copyright © American Institute of Chemical Engineers. All rights reserved.

434754 Wireless, Thermal Deactivation of Medical Device Infections Using an Iron Oxide Nanoparticle / Polymer Coating

Abstract:

Approximately 150,000 patients annually in the U.S. suffer from bacterial infections associated with implanted medical devices. The biofilms formed in these infections are notoriously difficult to treat, and despite decades of research on non-fouling surfaces and antimicrobial agents, most of these infected devices must be surgically explanted and replaced. Our work investigates an alternative treatment strategy using heat to thermally deactivate these infections *in situ* without the need for additional surgeries. Remote heating can be achieved using an iron oxide (magnetite) nanoparticle / polymer coating exposed to an alternating magnetic field (AMF) (i.e. magnetic induction heating). This application applies heat at a precise location, requiring a much larger power density than previous magnetic hyperthermia applications that have been investigated for decades in suspension form. In this work, we demonstrate the effects of several design parameters on the overall heating of magnetite coatings, specifically, iron concentration, coating thickness, the type of immobilizing polymer matrix, coating swelling, and orientation of the coating relative to the exposed AMF. Magnetite nanoparticles were immobilized in two, FDA-approved polymer matrices: (poly(styrene) (PS) and poly(vinyl alcohol) (PVA)). Although the power output of these films scaled proportionally with increasing iron loading, other parameters indicate the design of a coating to achieve uniform heating is much less straightforward. Characterization of magnetite-polymer films revealed heating depends primarily on the polymer matrix and orientation of the film relative to the applied AMF. Heating rates were observed up to 717 W/g Fe for hydrated PVA films positioned parallel to magnetic field lines versus only 100 W/g Fe for non-hydrated PVA films positioned perpendicular to the applied AMF. In addition, a separate, in vitro thermal model was built to quantify the power density that may be needed for bacterial deactivation on a surface inside the body. The maximum power density measured from all polymer, orientation, iron concentration, and coating thickness combinations was 7.5 W/cm^2 which exceeded the peak power requirement (4.5 W/cm^2) needed to achieve an 80°C surface temperature in 15 s beneath a tissue mimic heat sink. This work aims to eliminate a major liability of implanted medical devices and the billions of dollars spent on explantation, recovery, and re-implantation.

Copyright © American Institute of Chemical Engineers. All rights reserved.

434105 Controlling Atmospheric Pressure Plasma Devices for Biomedical Applications

Abstract:

Plasma biomedicine is an emerging field that seeks to apply cold atmospheric pressure plasmas (CAPs) for the improvement of health outcomes and the treatment of disease states. Recent research in this field has demonstrated the usefulness of these plasmas for disinfecting surfaces, inactivating viruses and prions, disrupting biofilms, improving wound healing, and even treating cancerous tumors, alone and in combination with traditional therapies. These biological effects are thought to be mediated by the complex combination of reactive neutrals, ions, radicals, UV radiation, electric fields, and dielectric heating produced by these discharges, where they can act by direct chemical interaction with the pathogen or tumor or by indirectly modulating blood flow, cellular redox state, cell cycle progression, biomolecular turnover rates, or immune system activity.

A wide range of plasma discharge geometries and power supply configurations have been suggested for these applications. In all cases, the operating performance of these plasmas is substantially affected by factors such as the ambient environmental conditions, the electrical properties of the treated surface, and the gap distance between the device and the substrate. CAP devices are inherently difficult to control due to the exponential dependence of ionization on temperature, positive feedback between ionization and spatial power dissipation, and the short mean free paths of gas molecules at ambient pressure. Repeatability of CAP device performance is often poor, even in laboratory settings.

In this presentation, we will discuss the design and validation of a control strategy for improving the performance of CAP devices.

Copyright © American Institute of Chemical Engineers. All rights reserved.

437560 Controlled Delivery of Growth Factors and Small Molecules for Peripheral Nerve Regeneration

Tuesday, November 10, 2015: 1:27 PM
250A (Salt Palace Convention Center)

Pratima Labroo[1], Himanshu Sant[2], Scott Ho[2], Bruce Gale[1], Jill E. Shea[3] and Jayant Agarwal[3], (1)Mechanical Engineering, University of Utah, Salt Lake City, UT, (2)University of Utah, Salt Lake City, UT, (3)Surgery, University of Utah, Salt Lake City, UT

Controlled local Delivery of growth factors and small molecules for peripheral nerve regeneration

Praima Labroo, Himanshu Sant, Scott Ho, Bruce Gale, Jill Shea, and Jayant Agarwal

Departments of Mechanical Engineering, Plastic Surgery, University of Utah, Salt Lake City, UT USA

Introduction

Autologous nerve grafts are the most commonly used graft to repair large nerve defects. Despite being the current "gold standard" autologous nerve grafts cause morbidity at the donor nerve site. Nerve conduits or tubes are a promising alternative to autografts. They act as guidance cues for the regenerating axons and allow for tension free bridging, without the need to harvest donor nerve. Separately, it has been shown that localized delivery of growth factors or small molecules can enhance axon growth and peripheral nerve regeneration. In this work we present the design of a novel drug delivery apparatus integrated with a PLGA based nerve guide conduit. This integrated device is designed to simplify the design process and provide increased versatility for releasing a variety of different growth factors or small molecules.

Experimental methods

The bioresorbable guidance conduits were produced using 75/25 poly-lactic-glycolic-acid (PLGA). PLGA was dissolved in acetone and ethanol and conduits were then formed. Diffusion hole(s) were drilled into the inner conduit by pulsing a laser cutter. Final assembly of mold-formed conduits and end caps was done using a solvent bonding process. Tests have been performed for release of dextran (Fluorescein tagged dextran), nerve growth factor (NGF) and tacrolimus. These drugs were loaded into the conduit reservoir separately and then the conduit was placed into a receiver chamber containing

DMEM F12 medium+10% FBS (fetal bovine serum) and 1% of Antimyotic solution at 37°C and 5% CO_2. This setup was used to replicate temperature and pH in physiological conditions. A series of sample collections were taken from the receiver chamber over specified time intervals and the chamber was flushed and filled with fresh media matrix each time. Fluorescence and ELISA readings were obtained and the data was analyzed to determine drug release kinetics. Collected samples were used to culture chick DRGs (dorsal root ganglia) for 72h to check the bioactivity of released samples. Drug delivering conduits loaded with NGF have been implanted in rats to test the growth across a transected nerve after 90 days.

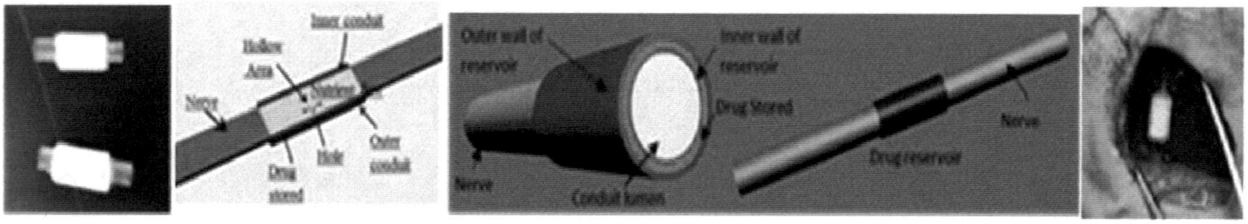

Figure 1. Photograph of the PLGA single-reservoir conduit prototype (left) (scale 5mm). Schematic diagram showing diffusion hole placement (center left). An axial view of the nerve and drug delivery device and schematic diagram of the drug delivery device placement across regenerating nerve (center right), drug delivering conduit implanted across cut sciatic nerve (right)

Results

The diffusion tests indicate that 150 μm hole allows for a sustainable release of NGF and dextran for much longer than 30 days. Five 50 μm holes were drilled for release of tacrolimus. Different hole dimensions were based on the dosage requirement and diffusion coefficient of each drug. The release was compared with the predicted outcome based on the simulations and collected samples showed bioactivity by enhancing growth of chick DRG axon length and density.

Histology of nerve and walking track data of rats is underway to confirm the success of drug delivering conduits as compared to autografts and no-drug conduits.

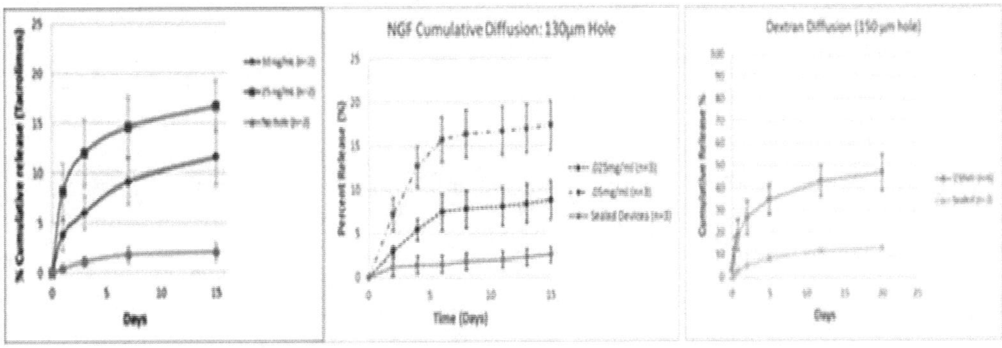

Figure 2. Cumulative release of tacrolimus (left) cumulative release of NGF (center), cumulative release of fl dextran (right) (legends indicate the release target of the device)

Collection period (no. of days)	DRG Axon length (μm)
0	707.0
2	945.5
3	807.2
4	780.6
7	971.0
25	486.6

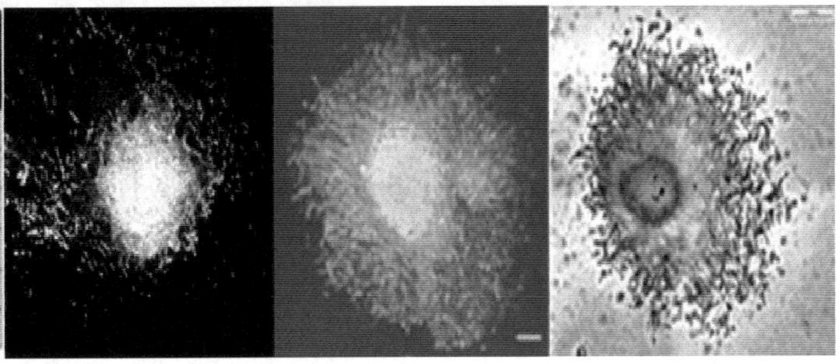

Figure 3. Table of DRG axon growth in presence of eluates of NGF released from drug delivery conduit at different days. The images showing DRG axon elongation after treatment with collected NGF sample (left) mage of DRG treated with 7-day collection of NGF releasing conduit (right) Image of DRG treated with 7-day collection of tacrolimus releasing conduit(right) (Scale 500 μm)

Conclusion

We have achieved controlled release of dextran, NGF and tacrolimus. The initial results with Sprague Dawley rat sciatic nerve model have shown improved nerve regeneration. Work is underway for a longer study using multiple growth factors and small molecules with rat and mouse models.

432533 Development of a Bioassay for Testing Inhibitors of Neutrophil Elastase

Abstract:

A balance in levels of alpha-1 antitrypsin (A1AT, a potent protease inhibitor) and neutrophil elastase (NE, a proteolytic enzyme) is believed to provide protection against development of chronic obstructive pulmonary disease (COPD), an umbrella term that includes emphysema and chronic bronchitis.

Therapeutics used for this condition are limited and generally only address the symptoms and not the underlying disease process. In an effort to better understand the kinetics of elastase inhibition with large and small molecule inhibitors, we have been developing assays that allow detection of degradation products of elastase-specific substrates. The goal is to develop an assay that considers options for reversibility or irreversibility that can then be used to test whether proposed, novel inhibitors of NE actually inhibit NE and if so what is the mode of inhibition (competitive, uncompetitive, noncompetitive, or mixed). In the assay, porcine pancreatic elastase (an analogue for NE) reacts with fluorescent elastin (a substrate for NE), and fluorescence is detected in the reaction by exciting the sample with light at 495 nm and detecting the emitted energy at 520 nm. The concentration of elastin tested has ranged from 0.2 to 4 mg/ml, and results reveal interesting profiles that are being further explored.

Copyright © American Institute of Chemical Engineers. All rights reserved.

(358f) Assessment and Control of ANTI-Microbial and ANTI-Inflammatory Responses of Macrophages to Different Titanium Nanomodifications

Bhardwaj, G., Northeastern University
Yazici, H., Northeastern University
Webster, T. J., Northeastern University
Originally presented on: 11/10/2015 14:05:00 - 14:24:00

ASSESSMENT AND CONTROL OF ANTI-MICROBIAL AND ANTI-INFLAMMATORY RESPONSES OF MACROPHAGES TO DIFFERENT TITANIUM NANOMODIFICATIONS

Garima Bhardwaj[1], Hilal Yazici [1],Thomas J. Webster[1].

[1]Northeastern University, Boston, MA-02115

Introduction: To monitor the immunological response of implants, the host response to materials must be determined prior to implantation. This response is highly dominated by macrophages which are the primary cells governing the immune response. Macrophages not only fuse to become multinucleated foreign body giant cells, but they also activate T lymphocytes by expressing co-stimulatory molecules (e.g., CD86 and CD80) and surface antigens (e.g., MHC II). T lymphocyte activation and giant cell formation have been observed on macrophages adherent to biomaterials and in retrieved adjacent tissues to failed implants. They are major effectors in the defense against bacterial infection and respond to interactions with gram-negative and gram-positive bacteria with a marked enhancement of their functional activities. Thus, the assessment and control of macrophage responses to various surface nano-modifications on titanium implants is of vital importance. The current study aims at modifying the surface of titanium by coating it with nanoscale hydroxyapatite, in a range of sizes, using electrophoretic deposition with DC current. Macrophages are then seeded onto this surface and their behavior observed concerning how to further improve the modification process.

Materials and Methods: Nanoscale hydroxyapatite was synthesized using a wet chemical synthesis process in 4 different sizes ranging from 110-170 nm, using Ca(NO3)24H2O, KH2PO4, distilled water and ammonia in an acid digestion vessel for hydrothermal treatment followed by drying in the oven. It was then coated onto a titanium mesh purchased from Alpha Aesar (Catalog no.7440-32-6) by electrophoretic deposition (EPD) with a DC current at 151 V for a minute each. Macrophages purchased from ATCC (RAW 264.7 (ATCC® TIB-71?)) were cultured using EMEM (ATCC® 30-2003?), 10% FBS (ATCC® SCRR-30-2020?) and a 1% penicillin-streptomycin solution (ATCC® 30-2300?). Cell adhesion and proliferation was observed using MTS assay for 1, 3, 5 and 7 days. Levels of TNF-α, IL-1, IL-6 and nitrite released by the macrophages were studied using PCR.

Bacterial assays were conducted using *Staphylococcus aureus* (ATCC® 29740?), Ampicillin resistant *E.coli* and *Pseudomonas aeruginosa* (ATCC® 39324?) strains of bacteria. 0.03% tryptic soy broth (TSB) and agar plates (Sigma-Aldrich) were used as the media. A small amount of bacteria was taken from the stock culture, streaked onto an agar plate, and then used as the stock plate for further experiments. Colonies were scraped off from the stock plate, added to 3 mL of 3% TSB and incubated at 37°C in humidified conditions under a 5% carbon dioxide atmosphere for 18 hours. A small amount (0.1 mL) of each sample was transferred to a few wells of a 96-well plate and absorption was measured at 562 nm using a plate reader. A value of 0.52 to 0.54 was obtained, indicating a density of 10^9 bacteria/mL. A dilution of 10^8 bacteria/mL was then prepared using 0.03% TSB. The samples were sterilized with 70% ethanol for 20 minutes, transferred into a 12-well plate, and rinsed once with PBS. They were then treated with 2 mL of the 10^8 bacteria/mL solution and incubated for 24 hours. The bacteria solution was removed and the samples were rinsed twice with PBS. They were transferred into 3 mL of PBS and sonicated for 10 minutes to create a first dilution (10^{-1}) then three subsequent dilutions (10^{-2}, 10^{-3}, and 10^{-4}) were created. Following this, 0.1 mL of each of the 10^{-3} and 10^{-4} dilutions were plated and incubated for 18 hours. The number of bacterial colonies formed on each sample was counted and using these values, the number of bacteria/mL was found. All experiments were conducted in triplicate and differences between means were determined using analysis of variance followed by Student's t-tests.

The particle size of the powders was determined using transmission electron microscopy (TEM) and the sample composition was established using x-ray diffraction (XRD). BET was used to measure the porosity of the powders. After coating, samples were characterized using scanning electron microscopy (SEM), atomic force microscopy (AFM) and contact angle analysis to confirm surface roughness and wettability of the coatings, all according to standard procedures.

Results and Discussion: Particle size and distribution was determined using TEM as shown in figure 1. After the samples were coated with the synthesized HA powders, SEM was used for surface characterization as shown in figure 2. This helped establish the presence of nano-crytalline features on the surface. Contact angle measurements were made using distilled water to test the hydrophobicity/ hydrophilicity of the surface, as shown in figure 3. The sample coated with HA by EPD showed complete wetting as compared to the hydrophobic behavior displayed by plain Ti. Figure 4 shows the result of confocal and fluorescence microscopy performed on the samples after seeding them with the 3 strains of bacteria and macrophages and fixing them. There was a decreased proliferation of these cells on the surface with reducing size of HA nanoparticles. Also, the change in the surface topography of the material affected the adhesion and proliferation of the macrophages leading to reduced activation of macrophages with increased nanometer roughness. The level of TNF-α, IL-1, IL-6 and nitrite released decreased with increasing hydrophilicity of surface but increased overall with time[1,2]. Bacterial activity decreased with smaller particle size of hydroxyapatite.

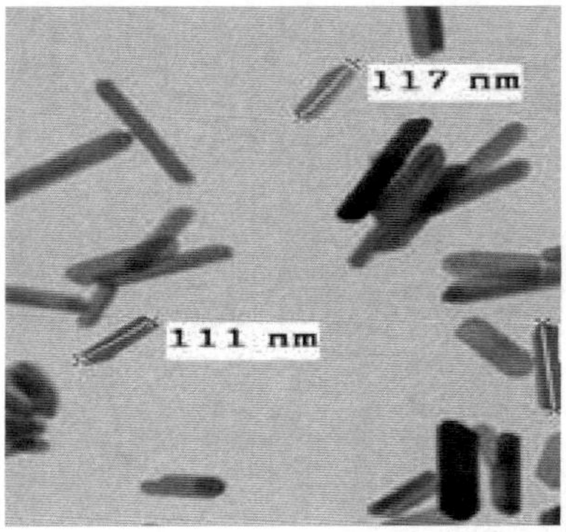

Figure 1: TEM image of the 110 nm HA powder. Scale bar - 100 nm.

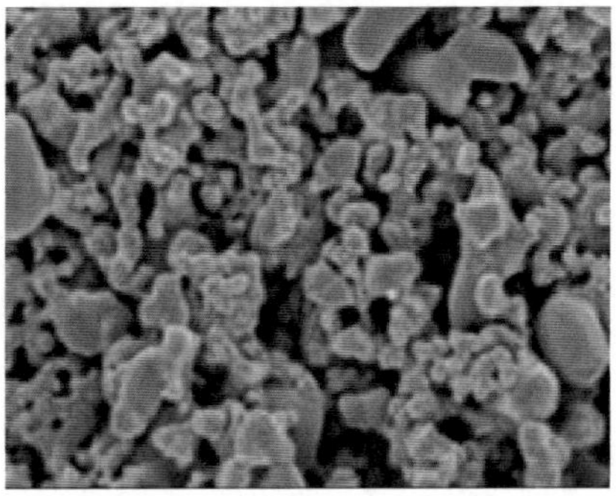

Figure 2: SEM image of the 110 nm HA powder coated onto the titanium mesh using EPD. Scale bar- 1μm

A

B

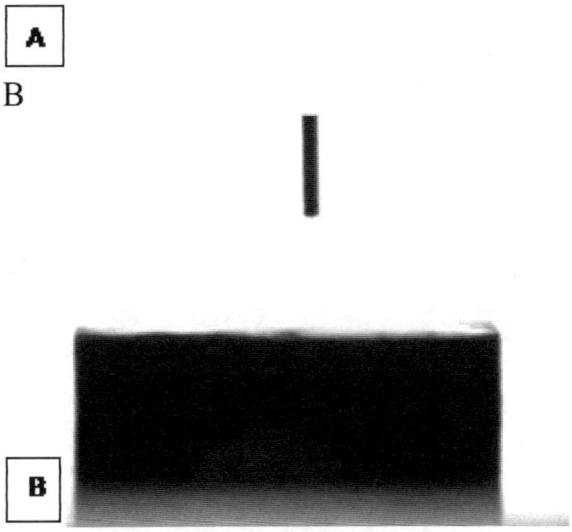

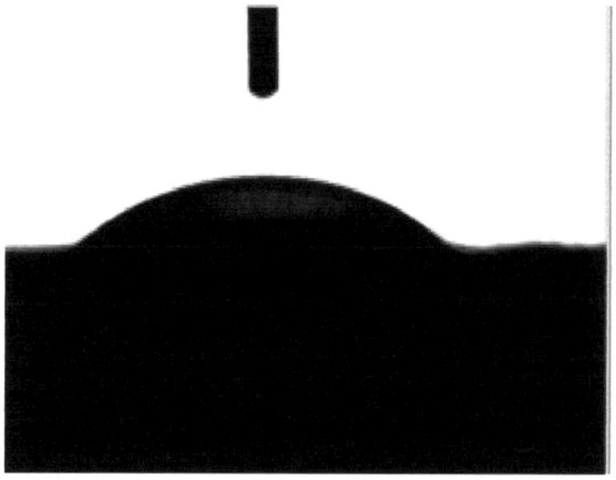

Figure 3: A) Contact angle image of the surface having 110 nm HA coated onto Ti using EPD. Image shows complete wetting of the surface establishing its highly hydrophilic nature. B) Contact angle image for the Plain Ti surface. Contact angle = 63.54°

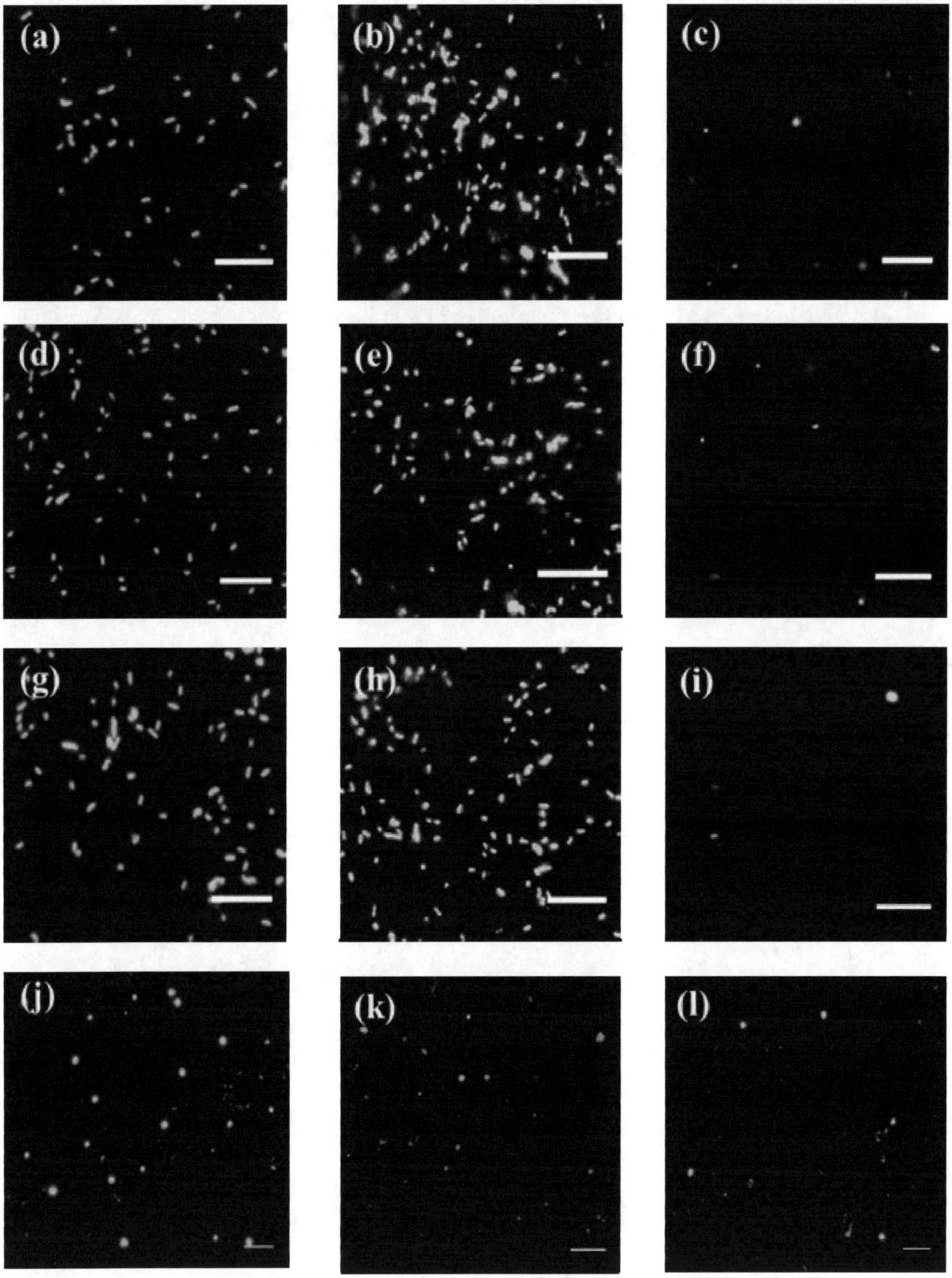

Figure 4: Confocal and fluorescent microscopy image showing bacterial and macrophage density on Plain Ti, Plasma Sprayed HA on Ti and Ti coated with 100 nm HA by EPD respectively. (a), (b), (c) represents S.aureus ; (d),(e), (f) represents P.aureginosa ; (g), (h), (i)

represents Ampicillin resistant E.coli. The scale bar for these images is 20 μm. (j),(k),(l) represents macrophage density on day 1 of culture. The scale bar for these images is 60 μm.

Conclusions: Controlling the size of the nanophase hydroxyapatite, the coating procedure and its parameters can help influence the fate of macrophage attachment, activation and functionality at the implant surface by reducing levels of TNF-α, IL-1, IL-6 and nitrite released with increased nanoscale topography. Bacterial activity is influenced by the size of the particles, the roughness, the hydrophobicity/hydrophilicity and the coating procedure used.

Acknowledgements: The authors would like to thank Northeastern University for funding.

References: 1.) Control of macrophage responses on hydrophobic and hydrophilic carbon nanostructures, Y. Chun et al, Carbon 49 (2011) 2092?2103. 2.) Reduced responses of macrophages on nanometer surface features of altered alumina crystalline phases, D. Khang et al, Acta Biomaterialia 6 (2010) 3864?3872.

437513 Copper Functionalized TiO2 Nanotube Sensors for Enhanced Rapid Glutathione Sensing

Abstract:

Oxidative stress is a physiological state characterized by excess levels of strong electron acceptors known as reactive oxygen species (ROS). Oxidative stress is associated with a number of human diseases, and may also result from exposure to hazardous chemicals. Because ROS are potentially damaging to cells, the human body uses endogenous antioxidants to control ROS concentration and maintain homeostasis. One of the most important endogenous antioxidants is a thiolated polypeptide known as glutathione. Glutathione in its reduced form (GSH) readily undergoes redox reactions that consume ROS and oxidize glutathione. Oxidized glutathione molecules thus produced link via covalent disulfide bridges, thereby forming glutathione dimers (GSSG). Hence the ratio of GSSG to GSH in various body fluids may be used as an indicator of oxidative stress. Here, we demonstrate proof-of-concept of a low-cost glutathione sensing platform that uses copper functionalized titanium dioxide nanotubes (Cu-TiO_2-NTs) to bind glutathione, and uses cyclic voltammetry (CV) to detect glutathione adsorption. Glutathione binding (GSH or GSSG) to the nanotubes is detected as a minimum in the current during a CV scan at -300 mV. The amplitude of this current minimum is larger for Cu-TiO_2 NTs than for bare TiO_2-NTs (3.18 times larger for GSH, and 1.71 times larger for GSSG). Our results demonstrate proof-of-concept for the use of copper functionalized TiO_2-NTs in a low-cost glutathione sensor.

Copyright © American Institute of Chemical Engineers. All rights reserved.

433552 High-Throughput Biomimetic Assay Designed to Quantify Antimalarial Efficacy

Abstract:

High-throughput Biomimetic Assay Designed to Quantify Antimalarial Efficacy

Megan A. Ketchum, Jeffrey D. Rimer, Peter G. Vekilov

Department of Chemical and Biomolecular Engineering, University of Houston, 4800 Calhoun Rd., Houston, TX 77204

Crystallization is associated with the pathogenesis of several pathological and infectious diseases in humans. In malaria, parasites utilize crystallization to prevent eradication by sequestering toxic hematin (released by the catabolization of hemoglobin) into innocuous crystals termed *hemozoin*. The sequestration of hematin into hemozoin is the target of numerous antimalarial drugs that inhibit crystallization, thereby increasing hematin supersaturation and eliminating the parasite from the body due to the cytotoxicity of hematin.[1] Parasite resistance to current antimalarials is spreading, fueling the demand for the development of new antimalarial drugs. Assays play an integral role in the discovery of new pharmaceutical drugs for diseases. High throughput assays are used by companies to test large libraries of drugs containing thousands of compounds to identify lead candidates. Here, we discuss a new method to screen for effective growth inhibitors of hematin crystallization. A relevant assay must mimic the parasite digestive vacuole (DV), which is a complex environment that is comprised of membrane interfaces, an acidic aqueous solution with pH 4.8 – 5.5, and lipids. While there is still debate regarding the medium (aqueous or organic) from which soluble hematin reaches the crystals and associates to them, we have provided convincing evidence that the lipid environment is crucial for crystal growth.[2, 3] To this end, we use a two-phase system of citric buffer saturated octanol (CBSO) to emulate physiological conditions in the DV. We designed an assay for quantifying hematin crystallization in the presence of putative growth modifiers to screen molecules as potential antimalarial drugs. This assay successfully determines IC
new drugs. The use of biomimetic solutions also provides a method of analyzing the fundamental mechanism(s) of drug-crystal interactions, which may prove to be an integral component of drug design that is capable of circumventing combinatorial approaches used to select and screen potential drugs.

References

1. Olafson, K.N., M.A. Ketchum, J.D. Rimer, and P.G. Vekilov, *Mechanisms of Hematin Crystallization and Inhibition by the Antimalarial Drug Chloroquine*. Proceedings of the National Academy of Sciences U S A, 2015. **112**: p. 4946-4951.

2. Ketchum, M.A., K.N. Olafson, E.V. Petrova, J.D. Rimer, and P.G. Vekilov, *Hematin crystallization from aqueous and organic solvents*. Journal of Chemical Physics, 2013. **139**(12).

3. Olafson, K.N., J.D. Rimer, and P.G. Vekilov, *Growth of Large Hematin Crystals in Biomimetic Solutions*. Crystal Growth & Design, 2014. **14**(5): p. 2123-2127.

Copyright © American Institute of Chemical Engineers. All rights reserved.

424089 Utility of DNA Capture Resins in the Development of Minimally Invasive Tools for Genetic Monitoring of Pancreatic Health

Abstract:

The diagnosis of pancreatic cancer carries an especially grim prognosis, with a median patient survival of less than 6 months, and an average 5-year survival of <5%. One reason for this poor outlook is the fact that pancreatic cancer is often diagnosed very late in the cancer development, resulting in the existence of distant metastases in more than one-half of patients at the time of diagnosis. There is a significant need for effective ways of detecting this cancer, early in its progression, which could lead to the evaluation and identification of better treatments for this deadly disease. Because ~95% of pancreatic cancers are exocrine in nature, pancreatic fluid represents a rich source of potential biomarkers that could lead to earlier diagnoses, and better clinical outcomes. However, the only method that is currently available for sampling pancreatic fluids is endoscopy, which is both time-consuming and highly invasive.

We are engineering a tool that would allow for the sampling and genetic screening of pancreatic fluids in a minimally invasive manner. We have identified materials that efficiently capture DNA from raw, untreated pancreatic fluid, thereby allowing it to be purified for downstream analysis. We have also performed time-course studies to evaluate fluctuations in pancreatic fluid DNA content upon secretin-stimulated secretion. These samples are now being utilized, in conjunction with next-generation sequencing, to identify resident genetic mutations. Here, I discuss these results and also provide data on our ability to detect rare mutant alleles within these fluid samples.

Copyright © American Institute of Chemical Engineers. All rights reserved.

422060 Fluorophore-Gold Nanoparticle Contrast Agent for Specific and Sensitive MMP-14 Detection

Abstract:

Matrix metalloproteinases (MMPs) are a family of endo-peptidases that contain zinc ions in their structures and are highly expressed in cancer. MMPs play crucial roles in the invasion, metastasis, proliferation, and development of cancer cells. MMP-14, one of the MMPs, is a membrane bound enzyme and is expressed at a high level in several aggressive cancer types, particularly in aggressive breast cancer (e.g. MDA-MB 231). It is known to promote the cancer cell detachment from the primary site, degrades and rebuilds the extracellular matrix (ECM), and stimulates tumor angiogenesis. MMP-14 also activates other MMPs, particularly proMMP-2, and is involved in the activation of extracellular signal-regulated kinase (ERK). Thus, specifically and sensitively detecting MMP-14 is important for early diagnosis of aggressive breast cancer, and specially designed, fluorophore mediated molecular sensing may be a good way to do so.

One way to increase the specificity of the fluorophore mediated molecular sensing is designing the contrast agent to emit its fluorescence only when this agent encounters the biomarker of interest. Increasing sensitivity of the contrast agent can be done by enhancing the fluorescent emission level per fluorophore.

Locally modulating fluorescent emission level of a fluorophore can be achieved by properly utilizing the electromagnetic (EM, plasmon) field generated by the gold nanoparticle (GNP) because GNP is small and can form a strong EM field around it. The EM field strength is highest on the GNP surface and exponentially decreases with the distance from the surface. Therefore, the fluorescent emission level can be controlled by appropriately varying the distance between the fluorophore and GNP. When the fluorophore is placed on the surface or very close to the GNP, its fluorescence is quenched significantly. When it is placed at an appropriate distance from the GNP, its fluorescence can be enhanced extensively. With this relationship between the EM field strength generated by a GNP and fluorescent emission level of a fluorophore, we have been developing a fluorophore contrast agent that is highly specific and sensitive for MMP-14 detection.

The fluorophore-GNP contrast agent for MMP-14 consists of a GNP, a near-infrared red (NIR) fluorophore Cypate (Cy), and spacers that link the GNP and Cy. An NIR fluorophore is chosen for a deep penetration to the tissue with a minimal background noise caused by the tissue auto-fluorescence, most of which are in the UV or visible wavelength. The GNP and Cy are linked by two different spacer types: The short spacer (SS) contains a substrate motif for MMP-14 and it is sufficiently short to place Cy close to GNP for the fluorescence quenching. When the agent encounters MMP-14, the SS is cleaved and Cy is released from the GNP, restoring its fluorescence. The long spacer (LS) is a stable, molecular sequence with a length for the Cy to be at the distance from a GNP for its maximum fluorescence. Ultimately, these two spacers will be conjugated to a single GNP together (also with an MMP-targeting molecule) so that the resulting contrast agent generates the Cy maximum fluorescence only when it encounters MMP-14.

In our process of optimizing the SS integrity, first, the length of the peptide sequence of SS was estimated by molecular simulation. Then, candidate peptide sequences possessing the MMP-14 substrate motif were conjugated to Cy (SS-Cy). Then the resulting SS-Cys were conjugated to the GNP (GNP-SS-Cy) and tested for their fluorescence levels and the cleavability of their SSs when they meet MMP-14. The fluorescence of the optimized GNP-SS-Cy was quenched up to 85% of its free form. When adding MMP-14, the fluorescence was restored up to 90%. We are currently testing the efficacy of the selected substrate motif in various cell lines. For the LS, we chose to use an amino acid sequence. The length of the optimal LS was first theoretically estimated and commercially available peptide candidates with potential length were selected. At a proper distance the fluorescence is estimated to be enhance up to 2.5 times of that of the free form. The GNPs conjugated with the Cy linked LS (GNP-LS-Cys) are currently tested for their fluorescence level.

The resulting products of the optimized Cy-SS/LS-GNPs with targeting moieties will be able not only to detect the cancer but also to diagnose cancer type. Lately, GNPs have used for various biomedical purposes for both diagnostic and therapeutic. In our study, the plasmon generated by the GNP was appropriately utilized to develop an optical contrast providing both enhanced specificity and sensitivity. Although this particular product is developed for MMP-14, the same concept can be used for the detection of any biomarkers that changes the spacer length, including enzymes, DNA, RNA biomarkers. In addition, GNPs are natural contrast agent for X-ray and/or CT, allowing contrast enhancement for two different imaging modalities.

Copyright © American Institute of Chemical Engineers. All rights reserved.

Design and Simulation of an Automated Rare Blood Cell Detector

Zhixi Qian[a], Eugene D. Boland[2], Paul W. Todd[2], and Thomas R. Hanley[1]
[1]Department of Chemical Engineering, Auburn University, AL 36849
[2]Techshot, Inc., 7200 Highway 150, Greenville, IN 47124

Abstract

Circulating tumor cells (CTC) plays an essential role in cancer's early detection and treatment. Meanwhile, the CTC's detection devices is limited in use due to its cost and complicity. To detect tumors at an early stage and improve patient's treatment quality and life expectancy, a point-of care cell detector can help in detecting CTC at clinics, nursing home or patient's home. Our goal is to design such a point-of-care CTC detector. The process of cell labeling and magnetic separation is simulated with the help of computational fluid dynamics (CFD) software. Models for fluid mixture as well as magnetic particles' track is set up and tested. We checked the relationship between magnetic ponderomotive force and magnet pole's size, and modified the magnetic particle velocity formula and a better description of magnetic particle's motion inside the magnetic trap. Consider a balance between and diffusion and sedimentation, we find an optimized value for the mixing process. Finally, a combine of theory calculation and laboratory experiments, the CTC detector is proved able to label and capture the minor CTCs in blood sample.

Key words: circulating tumor cells; CFD simulation; magnetic separation; cell labeling; static mixer

Introduction

In recent years there's an increasing demand of using circulating tumor cell (CTC) counts in cancer management. The primary objective of this research is to design, simulate and test an accurate, cost-effective, user-friendly point-of-care CTC detection device. The preliminary design includes three syringes for blood sample, buffer and reagents, a static mixer, a magnetic filter, a microfluidic "chip" and a waste receptor. The device is smaller and cheaper than traditional detection devices. Also, device operation is easier with only one operator required. The device can be used in clinics, nursing homes or residences thus allowing earlier detection of most fatal cancers, including pancreatic cancer, ovarian cancer and melanoma. Our recent research focused on the analysis and simulation of static mixer and the flow cytometer and a further analysis of factors affecting the fluid mixing.

The fluid dynamic analysis of the staged mixer was accomplished using the FLUENTTM analysis package, which performs modeling of fluid flow and tracking particles' movement in complex geometries. The simulation showed that inside the magnetic trap, the fluid performs a well-developed laminar flow. And it's estimated most magnetic particles will be captured at first pair of magnets.

An experiment was set up to test the magnetic trap's separation ability. Various conditions were tested, including flow rate, magnetic size and type. In addition, we tested the blood cell's sedimentation inside the magnetic trap. Then it's possible to optimize the device's flow rate. Hemacytometer and Hyperflux was used in analyzing particle patterns of the mixture.

Devices and methods

During the research task CFD software simulation and mathematic methods were applied. Computational fluid dynamic analysis of the channel design was accomplished using the FLUENTTM analysis package, which performs modeling of fluid flow and tracking particles' movement in complex geometries. With the help of ICEM CFDTM, a mesh file of the channel geometry was developed, and a laminar flow model was employed for calculation. For the simulation of magnetic particles, the channel was simplified to a 2-D rectangle shape. A discrete model was applied to simulate the particle movement. Tracking of particles movements was performed to show the particle trapping efficiency.

The trajectories of magnetic particles were simulated by choosing a set of coordinates within the channel and adding the flow and magnetophoretic velocity vectors as a function of position in the mesh. At each step in time the velocity is recalculated to account for simultaneous changes in streamline velocity, dy/dt, and magnetophoretic velocity, dx/dt.

To test the magnetic trap's capture efficiency, a hemacytometer is used in counting and calculation of magnetic particle captured. The red blood cell (RBC) and magnetic beads are put in sample tubes. After counting the origin concentration, the sample containing tumor cells is diluted and changes its concentration to a value closer to actual condition. Then the tumor cells will get labeled by magnet beads inside the static mixer. The mixture will then flow through the magnet trap, where tumor cells will stayed inside with free-moving magnet beads. The origin sample, the captured cell and escaped fluid will be analyzed in Hyperflux to check for its magnetic mobility and sizing information as well.

Magnet beads used in the testing are Dynabeads® Biotin Binder (InVitrogen/Dynal), the producer InVitrogen in Oslo, Norway and the catalog number is 11047. The magnet bead's magnetophoretic mobility range about 1.3~2.0*10-11m3/TAS and the concentration of beads is 4*108 beads/mL in the original product.

The Hyperfluxtm Velocimeter system is a particle analyzer and velocimeter that is used to measure the speed of movement of cells based on magnetophoretic mobility. It is driven by a calibrated magnetic field within the device. In addition to measured speed, the Hyperfluxtm also measures and generates 22 other characteristics about each identified particle including: size, sedimentation rate, shape, and intensity. The image sets created by the high-definition video camera are stored in memory can be used for processing and data evaluation. These image sets were captured on using the IKOVision program with specifically selected particles and operation conditions.

The device's magnetic separation ability was tested using known labeled target cells. The magnetic beads and cells were mixed in centrifuge tubes on a shaking bed for 20 minutes at 23 C. Thus all target cells are considered as labeled.

First we use a high concentration cell sample that will provide an indication of the magnetic trap's performance. It's expected that the device is able to keep a relative high and steady capture efficiency at a high tumor cell concentration. Then we dilute the cell suspension to the lowest practical concentration and check the separation patterns using hemacytometer counts and with the help of Hyperflux system. The system can check the diameter and mobility of particles, thus the difference of magnetically labeled cells and unlabeled cells can be understood.

Then we attach the static mixer upstream of the magnetic trap to test whether the system can successfully perform a test from labeling cells to separating tumor cells. After proving performance, whole blood is used in the final test, thus leading to a conclusion concerning potential clinical performance.

Results and discussion

At low cell concentration, the capture ratio cannot have a precise value owing to the low number of cells available in the hemacytometer, which was the rationale for the research at high cell counts described in the previous section. The starting and separated samples from these tests were subjected to detailed mobility analysis using the Hyperflux Velocimeter.

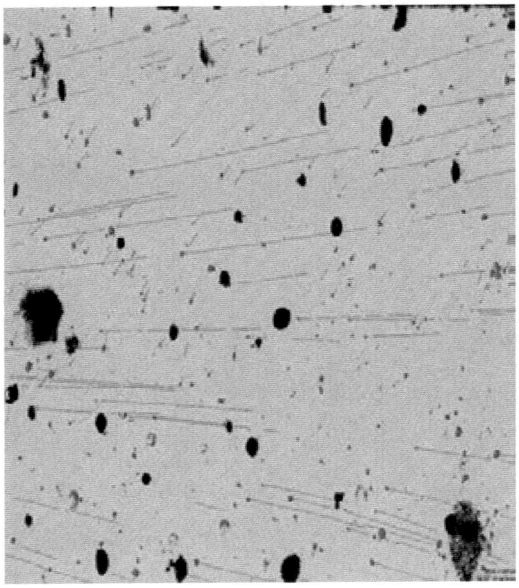

Figure above is the track screenshot for original sample. The red dots with long tracks are particles with high mobility, including cells labeled by magnet beads, and the red dots with short blue lines are particles with low mobility, which are not affected by the magnetic force.

n the starting sample, the ratio of underflow particles (those with zero or less velocity) is 29.1% of the analyzed tracks. In the 2-dimensional graph of diameter vs. mobility in Figure 5-9 there are points gathered around the y-axis (with a mobility less than 0.002), and at the right side of the xy-plot, the scattered dots represent labeled cells and magnetic beads which haven't labeled cells.

The track images for cells that were collected by the magnetic trap are shown below:

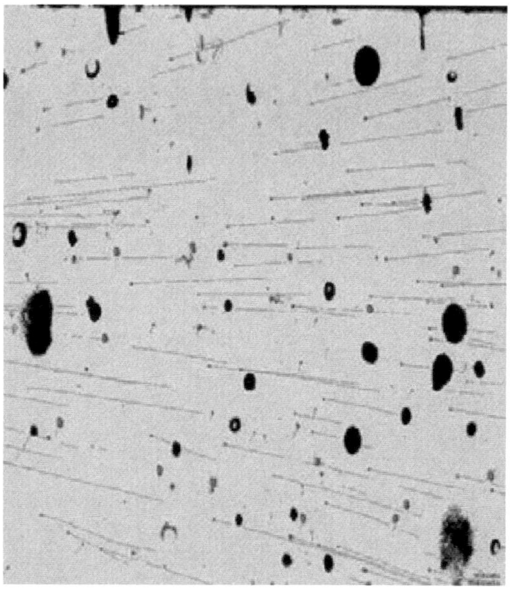

Compare with origin sample, less low mobility particles are tracked and more high mobility cells are observed.

And for the waste collected, the track is mainly from the particle with low mobility which is shown in fig below:

Fig 5-14: track images for waste of sample1

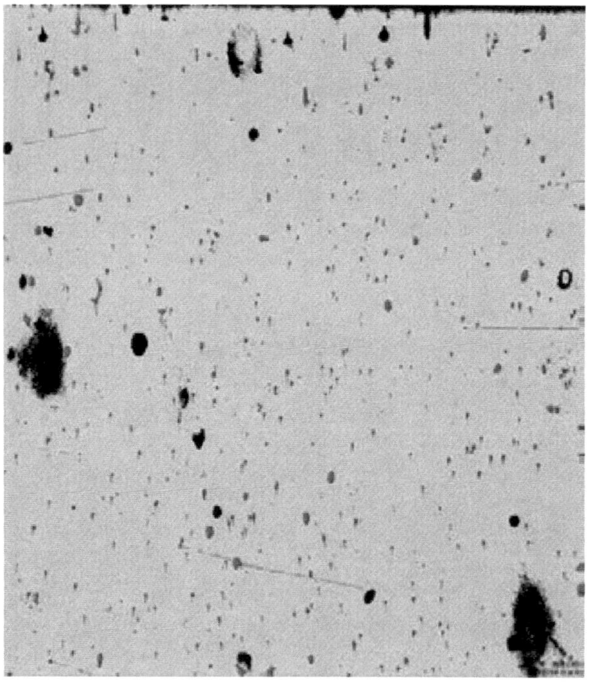

And the Hyperflux peak also showed a scattered peak in fig 5-15. Which comes mainly from the first peak in origin sample.

From the chart, there's a plenty of unlabeled particles in the part of waste, a combine of waste fluid and collected cell.

All these indicated the magnet trap's separation process. Three different types of particles are involved: particles with low mobility and varied diameter, which is unlabeled cells, they are mostly gathering in the waste after separation; particles with high mobility and low diameter, they are usually unattached magnet beads and will gather at collected sample. The labelled cells have a large diameter and sometimes even higher mobility as there're multiple magnet beads attached on the cells.

The result of Hyperflux tracking showed the magnet bead's mobility is between $1.3\sim2.0*10^{-11}m^3/TAS$, which is a lower value than in chapter 3, so the simulation result is different as well:

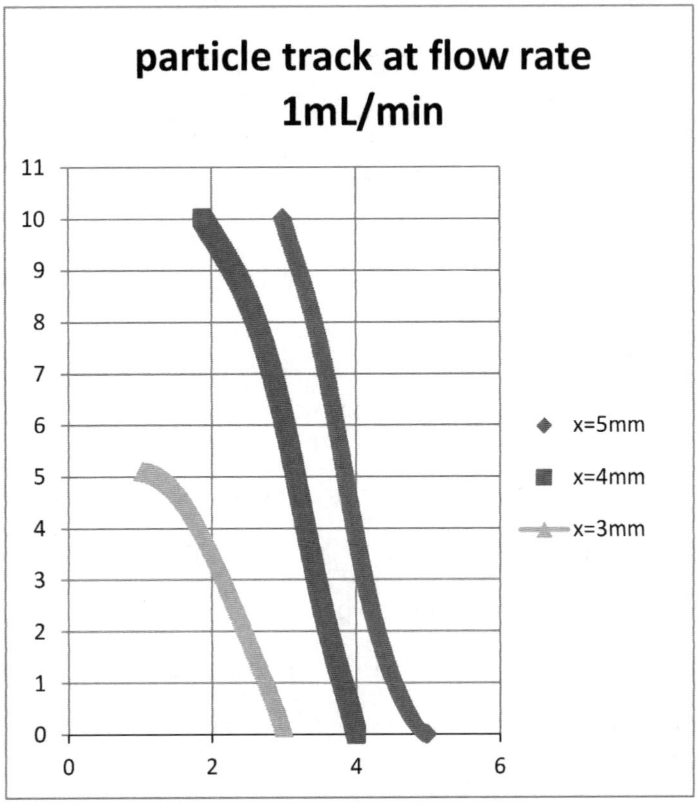

The first stage of beads can't capture all beads inside the trap, however, the beads which have a distance from magnet less than 2mm can be captured. After 1-stage magnet capture, those beads at x=5mm has moved to position of x=3mm, which will able to be captured at the second stage. So the 3-stage magnet ensures the capture ability.

Conclusion

The result of high concentration capture showed the device's ability in handling large amount of tumor cells. In blood sample, the tumor cell's concentration is less than 1000 cells/ml, so the test condition with 10^6~10^7 cells/ml is greatly exceeding the upper limit of the magnet trap. Still, the device can stably capture over 60% of labeled tumor cells. This proves the number of labeled cell device can capture is over 10^6 cells, which is greatly larger than the amount of cells actually in blood sample. And this showed the system is adequate in performing the task of cell separation.

The analysis result from the Hyperflux showed difference particles and their properties in the mixer. Unlabeled cells are varied in size and have a mobility around 0, free-moving magnet beads have a small size and high mobility, and the labelled cell have a larger size than free-moving magnet beads and the highest mobility. In the analyze report of magnet beads, we may see the origin sample contains all three types of particles. After separation. Most particles with low mobility is flowing through the separator and into the waste flow, the captured cells and free-moving magnet beads stay inside the device. The Hyperflux result proves the device can separate the labelled cells from unlabeled one with a good selectivity.

The result of magnetic bead's track showed difference type of magnetic beads can determine the performance of separation. A higher mobility results in a larger drag force performed on the cell. The multistage design proves necessary in supporting a wider range of magnet bead's choice. In this test, the labelled cells are collected within the first two sets of magnetic poles, indicating labelled cells labeled with lower magnet mobility can still be captured by the magnetic trap.

References

Afshar, R., Moser,Y, Lehnert, T and Gijs, M. A. 2011. Three-dimensional magnetic focusing of superparamagnetic beads for on-chip agglutination assays, Annul. Chem., 83(3) : 1022-1029.

Allard, W. J., Matera, J., Miller,M.C., Repollet, M., Connelly, M. C., Rao, C., Tibbe, A. G. J., Uhr J. W and Terstappen, L. W. M., 2004. Tumor cells circulate in the peripheral blood of all major carcinomas but not in healthy subjects or patients with nonmalignant diseases, Clin. Cancer Res.10: 6897-6904.

Cancer Data and Statistics, http://www.cdc.gov/cancer/dcpc/data/

Etzioni, R., Urban N. and Ramsey S., 2003. The case for early detection, Nature Reviews (Cancer), 3: 1-10.

Fan, T., Q. Zhao, J. J. Chen, W-T. Chen and M. L. Pearl, 2009. Clinical significance of circulating tumor cells detected by an invasion assay in peripheral blood of patients with ovarian cancer, *Gynecologic Oncology*, **112**, 185-191

Haun, J. B., Yoon, T-J. , Lee H. and Weissleder, R. 2010. Magnetic nanoparticle biosensors, Nanomedicine and Nanobiotechnology, 2: 291-304.

Hsieh, H. B., Marrinucci, D. , Bethel, K. , Curry, D. N. , Humphrey, M. , Krivacic, R. T. , Kroener, J. , Kroener, L., Ladanyi, A. , Lazarus, N, Kuhn, P. , Bruce R. H. and Nieva, J. 2006. High speed detection of circulating tumor cells, Biosensors and Bioelectronics, 21(10): 1893–1899.

434587 Biofilm Mitigation on Implanted Devices

Abstract:

Each year in the United States thousands of people have medical implants removed due to bacterial biofilm infections obtained during implantation. These biofilms are notoriously difficult to treat for a variety of reasons, including transport resistance from the extracellular polymeric substance, changes in gene regulation, and decreased metabolism from nutrient limitations. Despite decades of work on antifouling surfaces and antimicrobial agents, these infections typically require invasive explantation and re-implantation surgeries coupled with long-term antibiotic treatments. One promising new approach to address this problem is *in situ* thermal mitigation. In this study, *Pseudomonas aeruginosa* biofilms were subjected to temperature shocks at 50°C, 60°C, 70°C, and 80°C with controls run at 37°C for an exposure times of 1, 2, 5, 10, 15, 20, and 30 minutes. To guarantee instant, uniform thermal exposure these shocks were performed by immersion and removal from pre-heated wells, though our lab is currently developing magnetic/polymer composite coatings which can supply the thermal shock using an externally applied magnetic field. Following thermal shock, the remaining viable colony forming units (CFU) were quantified by serial dilution and plating of the homogenized biofilm. The resulting bacterial load in CFU/cm^2 fit the following equation:

$$\log(CFU/cm^2) = \log(CFU/cm^2)_0 - [0.079 + 0.044 \log(t)] * (T - 37)$$

where $\log(CFU/cm^2)_0$ is the initial bacterial load, t is the exposure time, and T is the thermal shock temperature. However, biofilms cultured under different conditions showed a difference in susceptibility to the heating, indicating that different areas of the body could lead to different needs for successful treatment. This work is the first step in a new approach to the goal of eliminating a major liability of implanted medical devices and the billions of dollars spent on explantation, recovery, and re-implantation.

Copyright © American Institute of Chemical Engineers. All rights reserved.

434042 Second Window Near Infrared Fluorescent Imaging for Deep Tissue Animal Imaging

Abstract:

Biomedical imaging is an integral part of clinical decision-making during screening, diagnosis, staging, therapy planning and guidance, treatment and real-time monitoring of patient response, because of its ability to provide morphological, structural, metabolic and functional information at various spatio-temporal scales of interest, while being a minimally invasive and highly targeted source of physiological evidence. The main clinical challenge, however, is to be able to image biological features with sufficient sensitivity for detection at the cellular level (in a complex tissue environment at deep penetration), which could then be used to identify and treat small tumor masses before the angiogenic switch growth phase, which is below the threshold of detection of most current imaging technologies. There are few main imaging modalities available clinically, classified according to the image contrast mechanism: X-ray (2D film imaging and computed tomography, CT), positron emission tomography (PET), single photon emission computed tomography (SPECT), magnetic resonance imaging (MRI), ultrasound (US), intravital microscopy (confocal, multiphoton), optical imaging, or some combination of these. Despite the significant advances in both the imaging instrumentation and algorithms for image processing, most of the aforementioned techniques suffer from poor sensitivity and/or resolution and/or depth of focus in the three spatial dimensions (3D). The most promising technique for high-resolution deep-tissue whole body imaging, using relatively safe molecular probes and excitation sources, at a reasonably low cost, appears to be optical imaging.

In recent years, there has been tremendous interest in the exploration of optical imaging *in vivo*, due to the development of near-infrared fluorescent probes, effective targeting agents, and custom-built imagers. Particularly, NIR light has been theoretically predicted to penetrate through ~ 10 cm of human tissue, with simulations suggesting a signal-to-noise (*S*/*N*) improvement of over 100-fold by imaging in the second near-infrared window (NIR-II: 900 - 1,400 nm) compared to the first window (NIR-I: 650 - 900 nm). Despite the recent development of various NIR-II fluorescent probes and custom built imagers based on InGaAs detectors, there are three significant challenges to imaging in the NIR-II wavelength domain with InGaAs detectors. The first challenge is the limited selection of fluorescent probes which emit in the NIR-II range favorable for medical imaging. Similarly to NIR-I or visible fluorescent imaging, only part of the fluorescent signal is utilized for contrast imaging to avoid spectral overlap with autofluorescence. The second challenge is to maximize the *S*/*N* ratio for high-sensitivity deep-tissue imaging. Due to the intrinsic low SNR of InGaAs detectors compared to Si detectors, strategies are required to depress SNR of InGaAs detectors by either imaging methodologies or data processing algorithms. The third challenge is diffuse light scattering by heterogeneous turbid biological media, imposing trade-off between depth and resolution.

In this work, we designed an optical system in NIR-II window that combines both hyperspectral and hyperdiffuse measuring. HSI resolves the first two challenges of SNR by providing information in frequency domain, and allows novel type of investigation (including pixel-wise spectral analysis and 3-D reconstruction) as well as improves result confidence. HDI resolves the third challenge, by not only excluding the emission scattering in the processed results, also presenting pixel-wise diffuse scattering information for contrast imaging and 3-D reconstruction. The hyperspectral and diffuse imaging system (900-1700 nm) has the capability to distinguish the optical signatures of the primary pump laser, background, various types of tissue autofluorescence, reporter fluorescence, as well as the diffuse scattering effect of the fluorescence signal upon transport through heterogeneous turbid optical media. Further by applying our imaging system to various tissue and whole-animal, we demonstrated that it is capable of (0) no background or autofluorescence information is required to be collected beforehand, and instead, these background and aufluorescence information is proved to be useful for both diagnostic and localization reasons; (1) detecting 1 mm sized particles up to 9 cm depth through a breast-tissue mimic phantom; (2) identifying spectral and diffuse signature of various tissues; (3) reconstructing 3-D fluorescent images based on newly interpreted information from HSC and HDC; (4) identifying important organs using label-free reflection mode hyperspectral imaging for feature location and registration. This study opens up exciting new possibilities for clinical translation of NIR-II imaging as a viable platform theranostic technology; for early diagnostics, as a real-time surgical assistance tool, and for monitoring patient response to therapies.

Copyright © American Institute of Chemical Engineers. All rights reserved.

433393 Specifically Tuned Light Activated Nano-Therapeutics for Selective Killing of Multi Drug Resistant Bacterial Strains

Abstract:

The recent surge of drug-resistant superbugs and shrinking antibiotic pipeline are serious challenges to global health. Here we present a novel nanoparticle-based therapy where the nanoparticles specifically target selective cell phenotypes based on their biochemical potentials. Using a specific light induced targeting mechanism, these light-activated reactive species (LARS) have the capability to target and kill specific-pathogens on illumination with visible light, without affecting the growth of other cells. While the growth of selected cells and the targeted pathogens is unaffected in dark, we demonstrate that LARS can eliminate a broad range of multi-drug resistant bacterial strains including extended spectrum beta lactamase *Escherichia coli*, carbapenem resistant *Klebsiella pneumoniae*, and Methicillin resistant *Staphylococcus aureus*. Furthermore, we demonstrate a potential therapeutic intervention with co-culture of *E. coli* and HEK293 cells, providing evidence that LARS is capable of removing an infection while leaving mammalian tissue intact. Finally, we designed different LARS species which are capable of promoting proliferation of targeted cells when induced with light. The ability to tune the biochemical potential of nanoparticles allows for selective cell phenotypes over a range of organisms, and has potential applications in antimicrobials and targeted organism-specific therapeutics.

Copyright © American Institute of Chemical Engineers. All rights reserved.

429133 Mathematical Modeling of Electrohydrodynamic Flow in Tumor Cells for Tumor Treating Fields (TTF) Therapy

Abstract:

Tumor treating fields (TTF) is a new type of cancer therapy that was recently approved by the Food and Drug Administration (FDA) in March 2011 for use in glioblastoma multiforme (GBM) patients that dramatically increases quality of life, while still being just as effective as traditional chemotherapy. TTF therapy uses alternating (AC) applied electrical fields to interfere with tumor cell division by disrupting mitosis via the movement of charged molecules within the cells, altering their internal structure. The processes of electrophoresis, electroosmosis and dielectrophoresis play a key role in the effectiveness of this therapy. Thus, here we will present a mathematical model that describes electrohydrodynamic flow within a single, dividing tumor cell during anaphase and cytokinesis by using idealized geometrical structures to describe the cell. Time dependent velocity profiles as a function of electrical field, cell phase and geometrical parameters indicate situations in which flow reversals can occur within the cells. These results can aid in ultimately predicting and optimizing tumor cell response to TTF therapy.

Copyright © American Institute of Chemical Engineers. All rights reserved.

433542 Mathematical Modeling of the Extracellular Matrix in Cancer Metastasis

Abstract:

Metastatic cancer accounts for 90% of cancer deaths in the U.S. Early in metastasis, cancer cells detach from the primary tumor site and move through the extracellular matrix (ECM) surrounding the tumor into a blood vessel from which the cancer can spread throughout the body. The ECM is a composite tissue composed primarily of a network of macromolecules including insoluble fibers and fluid-filled pores. Normal ECM serves as a physical barrier to the movement of cells. Cancer cells manipulate the ECM through cellular signaling pathways that activate enzymatic remodeling of the structure of the ECM to facilitate cell migration. ECM remodeling leads to a spatially heterogeneous ECM that changes over time. However, the mechanism of ECM remodeling and the impact of ECM remodeling on physical properties that influence metastatic cell migration are not well understood (Lu, Weaver, & Werb, 2012). Experimental approaches alone cannot decouple the complexities of the many interacting processes during ECM remodeling. Thus, there is a critical need for alternative approaches to understand how ECM composition and structure are modified during metastasis and to determine if ECM properties may be restored to non-metastatic values. Computational models considering simultaneous physiological, chemical, and biomechanical interactions can be developed to enhance quantitative understanding of ECM remodeling in cancer.

In this work, we present a computational model to of ECM remodeling in the early stages of cancer metastasis. The model includes the dynamic reaction and diffusion processes involving ECM-degrading matrix metalloproteinases (MMPs) and ECM-cross-linking lysyl-oxidase (LOX) and their impact on the collagen fibers in the ECM to predict ECM physical properties. This model improves upon existing models for the ECM in tumor metastasis by considering the transport variations due to the heterogeneous medium and the enzymatic reactions of both MMPs and LOX to both degrade and cross-link the ECM in tandem. The concentrations of the MMPs, LOX, and ECM basement membrane and interstitial collagen fibers are tracked with a mechanistic mathematical model that is solved computationally. The simulation results are assessed as to their ability to accurately predict stiffness changes due to ECM remodeling in the metastatic disease state as compared to published experimental data in the literature.

LU, P., WEAVER, V. M. & WERB, Z. 2012. The extracellular matrix: a dynamic niche in cancer progression. *J Cell Biol,* 196, 395-406.

Copyright © American Institute of Chemical Engineers. All rights reserved.

435011 Computational Model of Single Cell Transcriptional Regulation and Cellular Networks Driving Liver Regeneration Following Surgical Resection

Abstract:

Liver resection is used clinically as a treatment for hepatocellular carcinoma and as a technique to enable live liver transplant. In both cases, the resected liver initiates a highly orchestrated regeneration program that is expected to culminate in complete restoration of liver mass. Despite decades of study, our understanding of the regulatory mechanisms underlying this repair process remains incomplete. This lack of complete understanding limits the ability to intervene in cases where a resected liver fails to regenerate to the levels required to support normal physiological function. Nevertheless, the availability of clinical data and pathology samples from regenerating and non-regenerating cases present an opportunity to employ modeling approaches that integrate data from different functional scales to better understand the mechanisms underlying clinical resection responses. Our present study seeks to address the gap in mechanistic understanding of liver regeneration by using a combination of single-cell based transcriptional analysis and computational network modeling, which can then be integrated with clinical data to improve prognostic outcomes.

Initiation and progression of regeneration relies on the coordination of the responses of multiple cell types over time. A canonical description of the regeneration process after resection includes transient activation of the resident macrophages by the injury, followed by priming of hepatocytes to re-enter cell cycle, growth factor production by hepatic stellate cells and sinusoidal endothelial cells, proliferation of hepatocytes, proliferation of non-parenchymal cells, and termination of proliferative processes via remodeling the matrix microenvironment. Disruption of this coordinated sequence can lead to deficits in liver mass recovery, as clinically observed in chronic liver disease.

We developed a computational model that accounts for the multi-scale nature of liver regeneration by integrating physiological-scale interactions; activation phenotypes of hepatocytes and liver non-parenchymal cells; and molecular signaling networks. We explored a range of model dynamics to identify parameter sets that account for experimentally observed regeneration profiles of multiple animal models of liver disease as well as the human liver transplant case. Model simulations predicted that aberrant activation dynamics of hepatocytes and hepatic stellate cells are key factors suppressing liver regeneration in multiple disease contexts.

We tested these predictions in the alcoholic liver disease case by acquiring a new experimental data set on gene regulation in hundreds of single and pooled cells obtained through laser capture microdissection of liver tissue. We analyzed the high-dimensional single cell scale gene regulation data to identify distinct cell populations, and extrapolated from the single-cell scale to cell phenotype distribution in whole tissue. Our approach characterized the transcriptional state of individual cells in the *in vivo* regenerating liver at a level of detail not previously achievable. Characterization of transcriptional profiles revealed cell-type specific, functionally distinct states for hepatocytes as well as stellate cells in our data. Surprisingly, we found the two liver cell types distributed among all identified functional states irrespective of the experimental condition (baseline or regenerating, alcohol adapted or control). However, relative proportions of the cells distributed among the identified functional states changed based on regeneration stage as well as adaptation to chronic ethanol intake, consistent with predictions of the computational model. These results suggest that chronic alcohol use impairs liver regeneration via significant alteration of cellular networks that are critical to initiation and progression of liver repair following resection. Our integrated experimental and computational analysis points towards new avenues for therapeutic intervention based on renormalizing the cellular functional states for improving surgical outcomes.

Copyright © American Institute of Chemical Engineers. All rights reserved.

429373 Modeling the Dynamics of Neuroendocrine-Immune Interactions in Collagen-Induced Arthritis

Abstract:

Modeling the Dynamics of Neuroendocrine-Immune Interactions in Collagen-induced Arthritis

R. Rao[1], D. DuBois[2, 3], R. Almon[2, 3], W.J. Jusko[2] and I.P. Androulakis[1, 4]

[1]Department of Chemical and Biochemical Engineering, Rutgers, the State University of New Jersey, Piscataway, NJ, 08854, USA

[2]Department of Pharmaceutical Sciences, School of Pharmacy and Pharmaceutical Sciences, State University of New York at Buffalo, SUNY Buffalo, NY 14260, USA

[3]Department of Biological Sciences, State University of New York at Buffalo, SUNY Buffalo, NY 14260, USA

[4]Department of Biomedical Engineering, Rutgers, the State University of New Jersey, Piscataway, NJ, 08854, USA

Rheumatoid arthritis (RA) is a chronic inflammatory disease that affects about 1% of the population in the developed world, and is primarily characterized by synovial inflammation and cartilage deterioration, typically in the joints of the hands and feet (1). The hypothalamic-pituitary-adrenal (HPA) axis, a critical component of the neuroendocrine system, is thought to be an important regulator of disease activity. It has been hypothesized that adrenal insufficiency; the inability of the HPA axis to mount an adequate anti-inflammatory in response to ongoing chronic inflammation in RA might contribute to disease pathophysiology (2). In addition, circadian variation in clinical symptoms has been well documented in RA patients. Moreover, a disruption in the natural circadian rhythms of major mediators of the HPA axis, such as cortisol has been observed (3). A mechanistic understanding of the interplay between the circadian rhythms of the HPA axis and the immune system might provide useful insights into RA pathologies and optimization of treatment regimens.

Animal models of RA have been extensively used to study disease pathogenesis and in the discovery of anti-arthritic drugs (4). A rodent model of collagen induced arthritis (CIA), which involves immunization of the animal with an emulsion of the complete Freund's adjuvant and type II collagen, is considered to be the gold standard in vivo model for RA (5). In this work we discuss the development of a mathematical model that aims to evaluate the dynamic interactions between the HPA axis and pro-inflammatory cytokines, and their downstream effects on paw edema, a key disease end-point in CIA. In particular, we investigate the changes in circadian secretion of corticosterone (CST) and pro-inflammatory cytokines after induction of CIA, that ultimately result in circadian variability in paw edema. Moreover, we recapitulate the phenomenon of adrenal insufficiency that has been observed in both CIA as well as RA patients. These simulations will aid in the characterization of neuroendocrine-immune response in animal models of RA.

Our model builds on earlier work (6-9) and is comprised of three components; a central oscillatory compartment that simulates the circadian secretion of CST in the HPA axis, a peripheral compartment that captures the downstream effects of secreted CST on cytokine expression, and finally, a compartment that accounts for the disease endpoints, specifically, paw edema. The oscillations in the central compartment are generated due to the negative feedback between CST and corticotropin-releasing hormone (CRH)/adrenocorticotropic hormone (ACTH) using a modified Goodwin oscillator model that incorporates Michaelis-Menten kinetics to express the degradation terms in the hypothalamus, pituitary, and adrenals. The HPA axis is positively modulated by the expression of cytokines in the peripheral compartment, which in turn is balanced by the negative feedback of CST. This results in an interlinked positive and negative feedback circuit that enables the investigation of the dynamics of HPA axis-cytokine interactions upon the onset of inflammation. In this model, we assume that the severity of paw edema is in direct correlation to the expression of pro-inflammatory cytokines. Furthermore, we explicitly model the emergence of adrenal insufficiency in response to chronically elevated cytokines by postulating an indirect response mechanism that effectively results in stimulating the clearance of ACTH in the HPA axis. Disease progression is initiated by a delay in the onset of the increased pro-inflammatory cytokine expression by using a simple transduction model (6).

Model predictions demonstrate that elevated cytokine expression results in circadian variability in paw edema progression, reminiscent of the circadian rhythms of disease symptoms exhibited by RA patients. Notably, our model predicts that an increase in mean secretion of CST after the induction of the disease is accompanied by a decrease in the amplitude of the CST oscillation. This phenomenon persists even after the development of adrenal insufficiency and the consequent drop in mean CST secretion toward levels present prior to the onset of the disease. This suppression of circadian rhythm has been experimentally observed in a number of important neuroendocrine and HPA axis mediators during adjuvant induced arthritis, including; ACTH, CST and adrenocortical ornithine decarboxylase (10). Further analysis of possible alterations in the phase of oscillation of CST and the cytokines during disease progression is currently ongoing.

In conclusion, our model captures key features of chronic inflammation in animal models of RA. In doing so, it highlights the importance of accounting for circadian rhythms while studying the neuroendocrine-immune system interactions in RA.

References

1. McInnes IB, Schett G. The Pathogenesis of Rheumatoid Arthritis. N Engl J Med. 2011 Dec 8;365(23):2205–19.

2. Harbuz MS, Korendowych E, Jessop DS, Crown AL, Li pdfan S. L., Kirwan JR. Hypothalamo-pituitary-adrenal axis dysregulation in patients with rheumatoid arthritis after the dexamethasone/corticotrophin releasing factor test. J Endocrinol. 2003 Jul;178(1):55–60.

3. Cutolo M, Masi AT. Circadian Rhythms and Arthritis. Rheum Dis Clin N Am. 2005 Feb;31(1):115–29.

4. Kannan K, Ortmann RA, Kimpel D. Animal models of rheumatoid arthritis and their relevance to human disease. Pathophysiology. 2005 Oct;12(3):167–81.

5. Brand DD, Latham KA, Rosloniec EF. Collagen-induced arthritis. Nat Protoc. 2007;2(5):1269–75.

6. Earp JC, DuBois DC, Molano DS, Pyszczynski NA, Keller CE, Almon RR, et al. Modeling Corticosteroid Effects in a Rat Model of Rheumatoid Arthritis I: Mechanistic Disease Progression Model for the Time Course of Collagen-Induced Arthritis in Lewis Rats. J Pharmacol Exp Ther. 2008 Aug;326(2):532–45.

7. Mavroudis PD, Corbett SA, Calvano SE, Androulakis IP. Mathematical modeling of light-mediated HPA axis activity and downstream implications on the entrainment of peripheral clock genes. Physiol Genomics. 2014 Oct 15;46(20):766–78.

8. Sriram K, Rodriguez-Fernandez M, Doyle FJ. Modeling Cortisol Dynamics in the Neuro-endocrine Axis Distinguishes Normal, Depression, and Post-traumatic Stress Disorder (PTSD) in Humans. PLoS Comput Biol [Internet]. 2012 Feb [cited 2015 May 1];8(2). Available from: http://www.ncbi.nlm.nih.gov/pmc/articles/PMC3280965/

9. Ramakrishnan R, DuBois DC, Almon RR, Pyszczynski NA, Jusko WJ. Fifth-generation model for corticosteroid pharmacodynamics: application to steady-state receptor down-regulation and enzyme induction patterns during seven-day continuous infusion of methylprednisolone in rats. J Pharmacokinet Pharmacodyn. 2002 Feb;29(1):1–24.

10. Cardinali DP, Esquifino AI. Circadian disorganization in experimental arthritis. Neurosignals. 2003 Dec;12(6):267–82.

Copyright © American Institute of Chemical Engineers. All rights reserved.

431123 Development of a Hypertensive Ovine Model to Study Implantation of Autologous Arteries and Veins

Abstract:

Hypertension is a leading cardiovascular risk factor worldwide, playing a significant role in ischemic heart disease, stroke and atherosclerosis. Co-existence of systemic hypertension in patients undergoing infrainguinal bypass surgery was estimated at 70% and 68% of re-intervention cases in coronary artery grafting procedures were reported to be hypertensive. Venous hypertension also leads to chronic vein insufficiency leading to superficial vein incompetence, limiting the availability of autologous vein grafts. Here, we describe development of a hypertensive ovine model that was employed to test the performace of autologous arteries or veins when transplanted under clinically realistic conditions.

Our study involved the development of an ovine renal hypertension model utilizing the 2K1C model, where nephrectomy was performed on the left kidney and renal artery stenosis was performed on the right kidney after 3 weeks. Following the onset of hypertension, carotid artery grafts (CAG) or facial vein grafts (FVG) were implanted into carotid arteries of hypertensive (HT) and normotensive sheep (NT). Autologous carotid artery (CA) was used as a control. In addition, valuable real-time hemodynamic data was collected through instrumentation that was inserted during grafting. The instrumentation included a Doppler flow probe, ultrasonic crystals and in-line pressure catheters to measure flow rate, dynamic changes of graft diameter and arterial pressure.

With the 1K1C model, we were able to achieve elevated systolic as well as diastolic pressures, leading to increased Mean Arterial Pressure (MAP). Our in-vivo hemodynamics data revealed that in hypertensive animals the pulse wave velocity increased and the dynamic radial compliance was significantly reduced, indicating increased stiffness, a hallmark of hypertensive vessels. Surprisingly, both CAG and FVG showed significantly reduced mechanical properties as well as collagen and elastin content following grafting, albeit FVG were affected to a higher extent by HT. Smooth muscle cell remodeling, as well as vascular function was also more significantly affected by HT in FVG as compared to CAG. Interestingly, transplanted FVG into HT animals were infiltrated by pro-inflammatory macrophages, while in NT animals, transplanted FVG were infiltrated mostly by macrophages of an anti-inflammatory phenotype.

We present here a pre-clinical large animal hypertensive model, which allows for real-time hemodynamic data monitoring, to evaluate remodeling and performance of autologous vessels. Our study revealed that autologous arteries lose many of the vascular properties when grafted under NT conditions, but are not affected by HT as much as the veins. On the contrary, grafted veins showed constructive remodeling under NT, while under HT they performed poorly. This data has repercussions in choosing appropriate graftable vessels harvested from patients suffering from hypertension.

Copyright © American Institute of Chemical Engineers. All rights reserved.

Development of Modeling Approaches to Predict Effects of Facilitated Wound Closure on Scarring

Stephanie Jorgensen
Dr. Robby Sanders

Department of Chemical Engineering
Tennessee Technological University

TTU

1

Outline

- Overall Motivation
- Phases of Wound Healing and Types of Facilitated Closure
- Approach Taken Towards a Solution
- Development of a Model
- Conclusions

TTU

2

Motivation

- Millions of people are injured in a way requiring wound closure (via home remedies and/or suturing), and many have wounds that result in excessive scarring and complications.
  - ~3.1% of ER patients undergo suturing procedures each year[1].
  - Keloid and hypertrophic scars frequently occur in these patients (71% occurrence)[2].

Idea: To create a mathematical model to predict the type of wound closure that will result in minimal scarring

3

Phases of Wound Healing and Types of Facilitated Closure

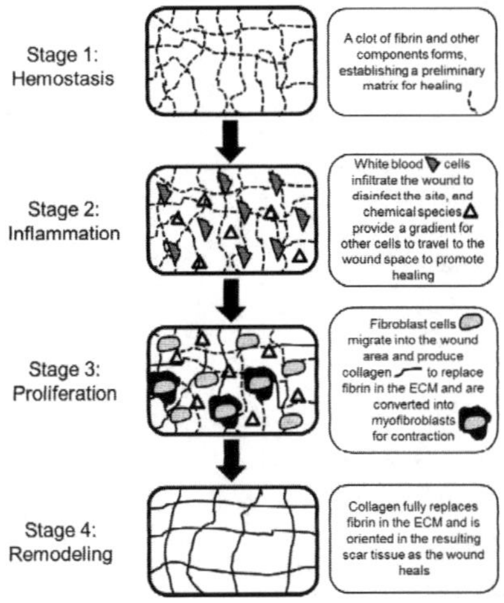

- Types of facilitated closure include
  - Sutures/stitches
  - Staples
  - Adhesives (tape, glue)

4

Approach: Collaborate with Healthcare Professionals

- Gain perspective of "needs" in healthcare through interaction with healthcare professionals
  - Developed a collaboration with faculty in the TTU School of Nursing
    - Facilitated access to Cookeville Regional Medical Center's ER to observe suturing and gain understanding about decisions relevant to suture material properties and stitching technique
    - *Lesson learned:* Techniques for closure/facilitated healing are based on experience (e.g., actual suturing in the clinic after limited training)

TTU

5

Approach: Conduct Comprehensive Literature Review

- Identify gaps between prior work by conducting a comprehensive literature review
  - Models included in 48 peer-reviewed articles (1981-2014) were classified using a two-tiered approach based on mathematical techniques applied and purposes of the models (including inputs and outputs).

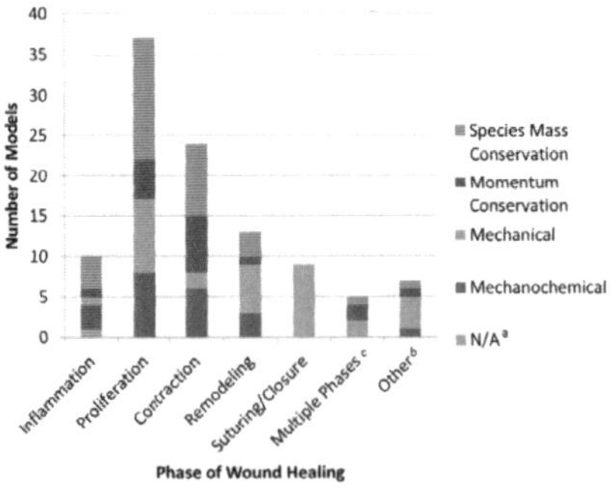

TTU

6

Approach: Conduct Comprehensive Literature Review

- A gap in the literature was identified.
  - Specifically, a lack of modeling of the events in inflammation and remodeling phases was observed, and few models were focused on effects of facilitated closure during wound healing.
- Opportunities for future mathematical modeling of wound healing exist with respect to these unexplored phases.

Goal: To create a mathematical model to predict the type of wound closure that will result in minimal scarring

TTU

7

Envisioning the Modeling Approach

- Determining the wound geometry and making assumptions for simplification (what's been done and considerations for my work)
  - Physical representations of the systems/domains utilized in typical models (from the comprehensive literature review) were identified, and the images were compared to typical wound domains that have been sutured (as illustrated in various medical resources).

TTU

8

1-D linear wound geometry	2-D, rectangular wound geometry	Circular wound geometry	Fusiform wound geometry
Has been used to investigate how concentration gradients affect fibroblast levels and tissue displacement during healing under normal and abnormal conditions	Has been used to determine how fiber alignment results from movement of cells within the wound and conclude contact guidance results in increased speed of healing	Has been used to determine how inhibition of lamellipod formation in cells and restriction of proliferation at wound edge affect wound closure	Has been used to determine regions of high and low stresses induced in wounds exhibiting facilitated closure

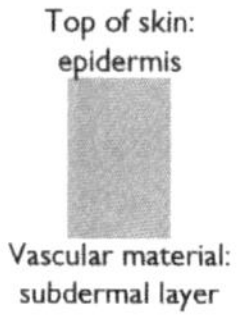

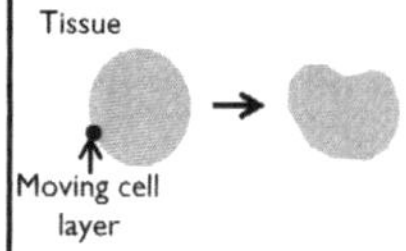

3

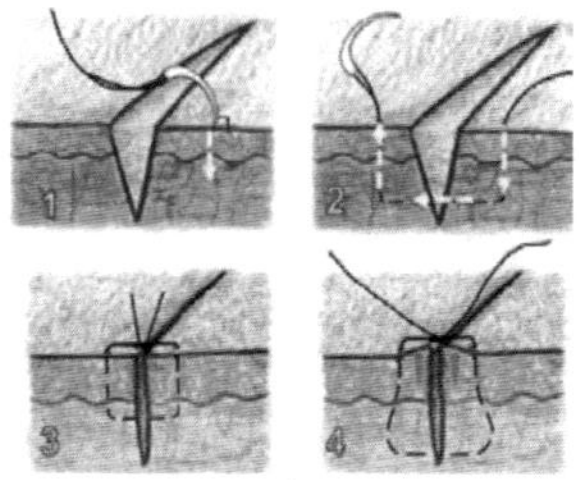

4

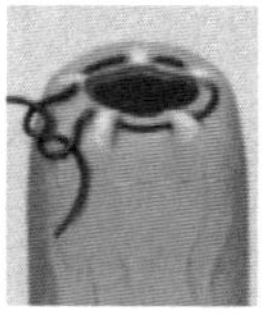

5

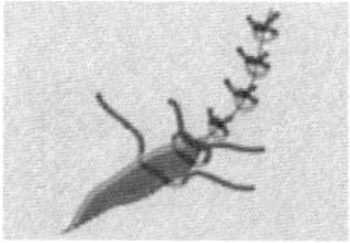

6

9

Envisioning the Modeling Approach

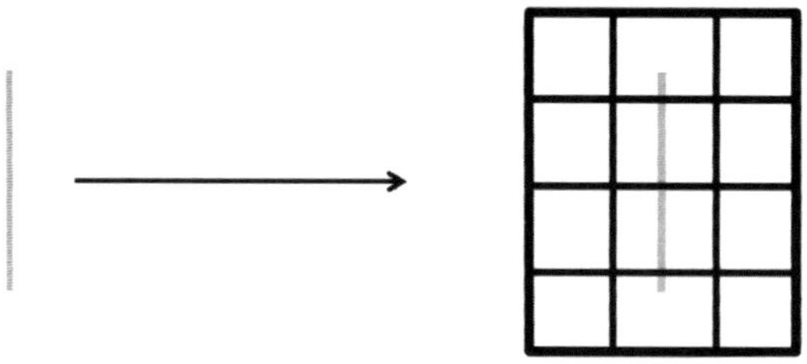

- Advantages:
  - Simple geometry for considerations of conservation equations
- Disadvantages:
  - Less physiologically relevant since wounds requiring suturing would be multi-dimensional

10

Envisioning the Modeling Approach

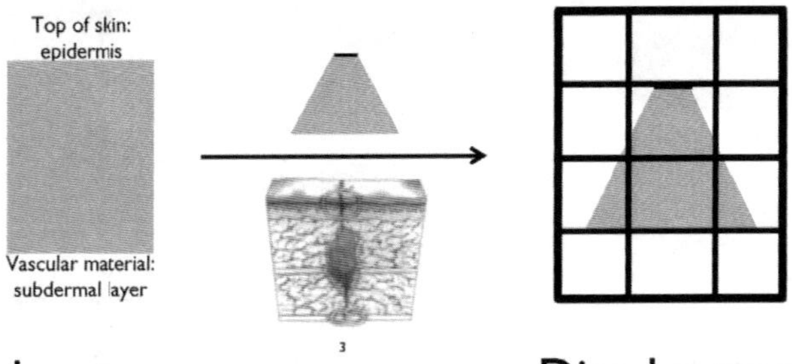

- Advantages:
  - Relatively simple geometry for considerations of conservation equations
- Disadvantages:
  - Not fully physiologically relevant as the "dead space" may not contain tissue elements associated with a typical healing response

11

Envisioning the Modeling Approach

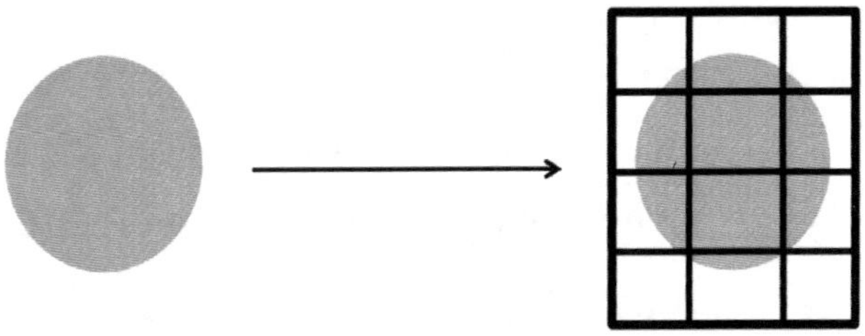

- Advantages:
  - Gives evidence of healing in different domains
- Disadvantages:
  - Geometry introduces many "trouble areas" in FEM—refinement of grid is necessary
  - Suture method presented (purse-string method) would pose a difficult boundary

12

Envisioning the Modeling Approach

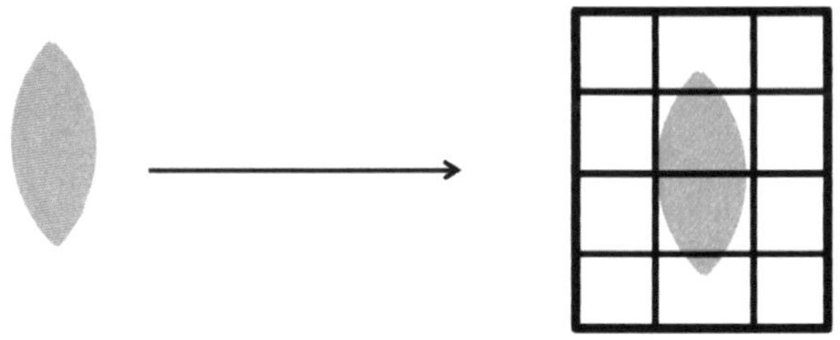

- Advantages:
  - Gives evidence of healing in different domains, and some wounds are intentionally made fusiform to achieve minimal scarring
- Disadvantages:
  - Complex for initial modeling efforts which is further complicated if consideration of subdermal forces is to be made

13

A Finite Element Approach

- Leveraging information from Larrabee and Galt[7]: we envision applying a force balance to wound boundaries using a triangular-type domain
- Determine the displacement of tissue along the wound boundary and within the healing wound as a result of facilitated closure

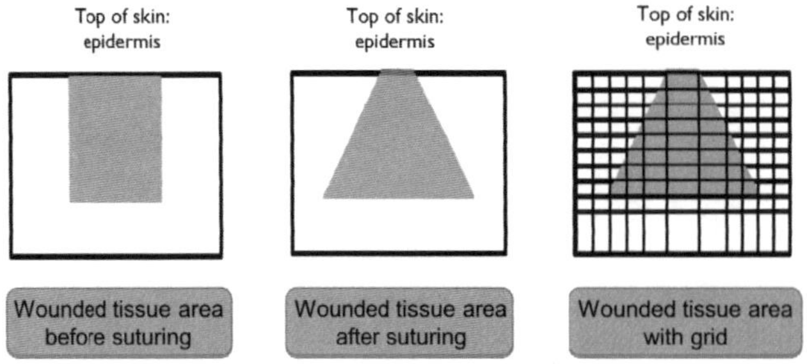

14

Conclusions

- Opportunities exist to better consider the effects of facilitated closure via suturing on the occurrences during the phases of wound healing
- Defining the system as it relates to facilitated closure/suturing is critical
- Next steps include generating a finite element mesh and/or model equations with appropriate boundary/initial conditions for the chosen domain

15

Citations

- 1 *National Hospital Ambulatory Medical Care Survey.* Centers for Disease Control and Prevention, 2010. Web. 25 Sept. 2014.
- 2 Li-Tsang, Cecilia W. P., Lau, Joy C. M., and Chan, Chetwyn C.H. "Prevalence of hypertrophic scar formation and its characteristics among the Chinese population." *Burns* 31 (2005): 610-616. Web. 15 Sept 2014.
- 3 Ethicon, Inc.,. *Wound Closure Manual.* Web. 24 Sept. 2015.
- 4 Brady, Mark. 'Optimal Suture Results: Slideshow'. 2015. Presentation.
- 5 Javed, Imran. 'Wound Management'. 2015. Presentation.
- 6 Brady, Mark. 'Optimal Suture Results: Slideshow'. 2015. Presentation.
- 7 Larrabee, W. F., and J. A. Galt. 1986. A finite element model of skin deformation, III. The finite element model. *The Laryngoscope.* **96**, 413-418.

16

Acknowledgements

- TTU Office of Research, Department of Chemical Engineering
- Research Committee: Dr. Robby Sanders, Dr. Pedro Arce, Dr. Jeffrey Rice, Dr. Jennifer Pascal, Dr. Melissa Geist, Dr. Yung-way Liu
- Sponsor: Chemical Engineers in Medicine
- Co-Sponsor: Computing Systems and Technology Division
- Chair: David Shaw
- Co-chair: Charles Reid

17

Questions?

18

A collision model of erythrocyte aggregates in shear flow

S.K. Ahuja

Abstract:

Platelet (thrombocyte) aggregation is one of the first physiological reactions to vessel wall injury. Aggregation in platelets and neutrophils (PMNs) is relevant to the progression of thrombosis, acute myocardial infarction, or unstable angina. Platelet aggregation at sites of vascular injury is central to the arrest of bleeding and for subsequent vascular repair; Excessive thrombocyte aggregation response at sites of atherosclerotic plaque rupture can lead to the development of vascular occlusive thrombi. In order to prevent it from happening, platelets move to the vessel wall and collide with either the damaged site or previously adhered platelets. Hemostasis is affected by erythrocytes when blood viscosity is increased through increase in concentration of erythrocytes or aggregation of erythrocyte takes place or when erythrocyte structure changes on deformation.

Erythrocyte (Red Blood Cell) aggregation is large at low shear rates and as shear rates increases aggregates break up. Bridging model and depletion model are used to analyze thrombus, clot formation. The adhesion of erythrocytes to each other and of erythrocyte to thrombocytes can take place. In the bridging model, it is assumed that fibrinogen or dextran molecules non-specifically adsorb onto the cell membrane and form a "bridge" to the adjacent cell. In contrast, the depletion model proposes the opposite. In this model, aggregation occurs as the concentration of macromolecules near an erythrocyte surfaces in close proximity is depleted compared to the concentration of the bulk phase, resulting in a net depletion force. Erythrocyte aggregation depends on their elastic properties. A model is proposed which considers collisions of thrombocyte with another thrombocyte or with an erythrocyte. Factors such as sizes of thrombocyte and erythrocyte, concentration of erythrocyte, shear rates in the thrombocyte to thrombocyte adhesion and thrombocyte to erythrocyte adhesion are discussed and compared with available experimental results.

Introduction: Blood clot (thrombi) formation is a complex biological process involving an extensive system of biochemical (coagulation) reactions, platelet hydrodynamics, platelet–platelet and platelet–blood vessel wall interactions leading to ligand–receptor adhesion bond formation and platelet activation. Platelet adhesion to the vessel wall is one of the first events associated with formation of haemostatic clots and pathological thrombi. Damage or alteration of a blood vessel lining can result in activation of flowing platelets and their subsequent aggregation at sites of vascular injury. The ability of platelets to tether to and translocate on injured vascular endothelium relies on the interaction between the platelet glycoprotein receptor Ib α (GPIb α) and the A1 domain of vonWillebrand factor (vWF-A1) [1]

Platelet activation and homotypic aggregation induced by high shear stress or by chemical agonists (e.g., ADP, collagen, or thrombin) are involved in atherosclerosis and accompanying thrombosis cell aggregation and disaggregation in a linear shear flow. Platelet adhesion to collagen via blood glycoprotein vonWillebrand Factor (vWF) is important for thrombosis initiation at high shear. VWF is shear dependent that transitions from a globular to elongated conformation under high shear rate, enhancing absorption onto a collagen surface and platelet adhesion. A collision model of shear induced cell aggregation and disaggregation is governed by receptor–ligand binding. Proaggregation forces are generated by chemical affinity of cellular receptors and their ligands. Anti-aggregation forces are generated by elastic collision, electrostatic barrier and osmotic pressure It was revealed that platelet aggregation rate does not simply increase with blood-flow velocity; rather, it exhibits a maximum with a subsequent decrease due to the growing hydrodynamic forces that inhibit platelet adhesion The combination of hydrodynamic features of microvasculature and nonlinear shear-dependent platelet aggregation rate may, in principle, stop the growth of a thrombus. [2,3].

Erythrocyte form aggregates in rouleaux formation with macromolecules (fibrinogen or dextran) present in plasma. The first step is the rapid formation of unstable catch–slip bonds which slow platelets and cause platelet flipping along the damaged surface. The platelet receptor component, GPIb α, forms transient bonds with the vWF exposed at the injury site. Transient tethering and subsequent flipping (or rolling) and pausing of platelets on the vessel surface result in stable bonds slowly between platelet receptors and ligands (often integrin αIIbβ3 binding with Vwf or fibrinogen) bound to the damaged wall or the surface of the thrombus. As the blood clot grows, platelet–platelet interaction becomes one of the major factors determining clot growth rate and integrity as platelets expose GPIIbIIIa receptors which permit platelet–platelet adhesion via fibrinogen. Adhesion of platelets to the injured surface is also affected by shear rates of the flow. At high shear, platelet integrin $\alpha2\beta1$ and GPVI receptors are not sufficient to initiate binding to collagen, and binding of the GPIb α receptor to vWF immobilized on collagen becomes essential in platelet adhesion[3]. *The stiffness of the platelet not only determines the shape and morphology of the clot, but also affects clot mechanical properties, as platelet stiffness determines cell shape when exposed to various flow conditions and contact interaction with other cells and blood vessel wall.* This will affect the number of receptor–ligand pairs in platelet–platelet and platelet–substrate interactions. Platelet stiffness is also an important property reflecting platelet functioning, because it reorganizes its structure during activation or as a response to physiological or pathological conditions [4].

Roll of Erythrocyte in blood clot::Thrombus structure has two distinct regions: a densely packed “core” nearest the injury site (composed of highly activated platelets that have undergone αgranule exocytosis and phosphatidylserine exposure) and a loose “shell” overlaying the core composed of less activated platelets that have retained a discoid shape The thrombus shell consists of loosely associated platelets that are subject to continual removal by flow forces. Within the thrombus core, platelet agonists may become enriched since molecular transit may be reduced by the core having restricted pores. The presence of erythrocytes in contracted blood clots has been recognized and sometimes exploited; for example, during the time of medical bloodletting, the size of the contracted clot from blood removed from the patient was used as a measure of erythrocyte mass to determine when the procedure should cease. Erythrocytes are uniquely deformable cells with shapes determined by their geometric and material properties that regularly undergo extensive, reversible shape changes under the influence of intravascular fluid forces and while passing through vessels in the microcirculation smaller than their diameter. However, the effects on erythrocytes of the forces generated by clot contraction and their impact on hemostasis have not heretofore been investigated [5].

There are possible mechanisms through which erythrocytes influence the balance of thrombosis and bleeding. Erythrocytosis can be caused by diseases affecting the bone marrow (primary erythrocytosis, such as polycythemia vera) or by disease and environmental conditions that affect blood oxygen saturation (secondary erythrocytosis). In secondary erythrocytosis, the bone marrow produces more erythrocytes to counterbalance the decrease in oxygen saturation. An increase in hematocrit results in an increase in viscosity that decreases the blood flow and can lead to the development of a thrombus. In addition, an elevated hematocrit promotes the transport of platelets and coagulation factors toward the vessel wall, thereby increasing collisions of platelets with the vasculature and with themselves. In a laminar blood flow, the highly deformable and biconcave shaped erythrocytes push platelets to the peripheral, and they themselves migrate to the center of the flow. This phenomenon is also called axial migration. The axial migration of erythrocytes causes platelets to be concentrated on the edge of the vessel wall. Erythrocytes are able to aggregate when suspended in aqueous solutions containing large plasma proteins (e.g., fibrinogen) or polymers (e.g., high-molecular weight dextran). At stasis or when fluid shear rates are sufficiently low, erythrocytes tend to form regular, linear, face-to-face structures (i.e., rouleaux), or three-dimensional aggregates that are difficult to disperse. Mechanisms involved in erythrocyte aggregation are based on two models.. The bridging model proposes that erythrocyte aggregation occurs when bridging forces generated from the adsorption of macromolecules onto adjacent cell surfaces exceed disaggregation forces generated from electrostatic repulsion, membrane strain, and mechanical shearing [6,7].

The other model is the depletion model where erythrocyte aggregation occurs because of a lower localized protein or polymer concentration near the cell surface compared with the suspending medium leading to an osmotic gradient and a depletion interaction. Although it is not yet possible to decide the superiority between these two mechanisms several studies favoring the depletion model have recently been published. In large blood vessels under low shear conditions, where blood can be regarded as a continuous fluid, enhanced erythrocyte aggregation increases blood viscosity and overall hydrodynamic resistance. In the microvasculature, however, enhanced erythrocyte aggregation increases axial migration of erythrocytes, resulting in an increased platelet concentration near the vessel wall, as well as a decreased local viscosity and reduced frictional resistance.

In addition, axial accumulation of erythrocytes also promotes plasma skimming and the Fahraeus effect, which in turn may result in decreased hematocrit values, thereby reducing blood viscosity and flow resistance. This reduction in local viscosity and consequent reduction in wall shear stress leads to a lower local nitric oxide bioavailability. Nitric oxide is an important substance in maintaining the quiescent status of endothelial cells and platelets. The decrease in local nitric oxide may promote platelet activation and tilt the hemostatic balance. Experimental studies indicate that lateral platelet movement is caused by continuous rebounding collisions between erythrocytes. These net collisions produce shear-induced diffusion (dispersion), which locally mixes the plasma and platelets near erythrocytes [8, 9, 10].. In a mesoscopic approach employing the immersed boundary method was used to simulate the motion of RBCs, modelled as deformable capsules, flowing through two-dimensional channels 20–300 mm in height. The results showed migration of RBCs perpendicular to the wall and formation of a cell-free layer[11].

In the immersed boundary method was also employed to model RBC migration in two dimensional channels of 6–12 mm in height; cell migration from the vessel wall was also observed. In the dissipative particle dynamics method was applied to blood flow in microtubes ranging from 10 to 40 mm in diameter, and cell migration away from the wall to the tube center was reported A different approach is represented by continuum models, which are able to simulate larger domains for longer time scales. An example of this approach can be found in where blood flow in vessels with diameter ranging from 40 to 100 mm was investigated using the shear-induced migration model proposed in. Particle migration away from the wall was also observed in this case. RBC distribution influences blood viscosity, and therefore the velocity profile, and also a series of biological activities such as oxygen distribution in the lumen and the scavenging of nitric oxide (NO) by haemoglobin in the RBCs[12]..

Erythrocyte aggregation: Platelets first became activated at locations where shear rates are greater than the critical shear rate (critical shear regions) of the first and second bends. Thrombus first grows along the wall in the downstream direction and then later in the upstream direction. Collision with erythrocyte with thrombus results in thrombus to roll along the wall. RBC collisions caused thrombi to continually roll along the wall and exit, The collision rates are consistent with the observation that the smaller molecules are more dynamic, and, therefore, have higher collision rates than the larger molecules. In a single layer of platelets that could contain many platelets. The colloid particle population is distributed by mass (or size) in a cluster mass distribution (CMD). The CMD evolves according to a population balance equation (PBE) which satisfies the principle of mass conservation. Smoluchowski was the first to apply statistical mechanics for aggregation of colloid particles.[13,14]. In processes where hydrodynamic forces are dominant, fragmentation may occur. Simultaneous aggregation and fragmentation by fluid shear is encountered in a number of processes. Aggregate breakup occurs by two mechanisms: erosion and splitting. Erosion is the detachment of small particles from surface, of aggregate and splitting is the breakup into pieces of size comparable to the parent aggregate. Competition between attractive van der Waals-Hamaker and repulsive double-layer forces determines the stability or instability of many colloidal systems. The interaction between colloidal particles can be modeled by the Derjaguin-Landau-Verwey-Overbeek (DLVO theory) where DLVO potential, which is the sum of the screened electrostatic repulsion between their double-layers and of the van der Waals-Hamaker attraction [15]. Shear-induced aggregation and Brownian aggregation are two mechanisms that exist simultaneously in sheared system.

Blood viscosity, is affected by various factors such as hematocrit, erythrocyte deformability and aggregation, osmolality, and plasma proteins. Rheological parameters of blood that have the potential to indicate blood circulation adequacy: yield shear stress (YSS) and disaggregating shear stress (DSS). YSS and DSS reflect the strength of red blood cell (RBC) aggregation in suspension under static and dynamic conditions, respectively. YSS, defined as the critical stress to disperse RBC aggregates under static conditions, was found to be dependent upon hematocrit, fibrinogen, and red cell deformability, but not temperature. DSS, defined as the minimum shear stress to disperse RBC aggregates under dynamic conditions, is dependent upon fibrinogen, red cell deformability, and temperature but not hematocrit[16].

Hydrodynamic stresses in elongation or shear in either laminar or turbulent regime can affect dispersion stability and breakage of colloidal aggregates in fractal or non-fractal morphology [16]. Balance between clustering (structure-formation) and breakup (structure-failure) is critical enabler in analyzing structural origin of the rheological properties of complex fluids.

$$\frac{dn_i}{dt}=\frac{1}{2}\sum_{j=1}^{i-1}k^{(a)}(i-j,j)n_{i-j}n_j-\sum_{j=1}^{\infty}k^{(a)}(i,j)n_in_j-k^{(b)}(i)n_i$$

$$+\sum_{j=i+1}^{\infty}\beta(i,j)k^{(b)}(j)n_j,$$

--------------------------------(1)

where the subscripts i and j refer to different size classes, n_i is the number of aggregates having i monomers per unit volume, and $\beta(i, j)$ is the fragment size distribution[26] (here binary fragmentation). The kernels $k^{(a)}$ and $k^{(b)}$ depend on the type of flow

$$k^{(b)}(i)=\dot{\gamma}\exp\left(-\frac{2F_c}{5\pi\rho_i^2\eta\dot{\gamma}}\right)$$

On application of the analytical aggregation kernels for simple laminar shear (orthokinetic aggregation), and thermal motion (perikinetic aggregation) to model the coagulation kinetics of Brownian particles under laminar shear flow and superimposing the two processes gives the aggregation kernel as

$$k^{(a)}(i,j)=c_o\left(\frac{4}{3}\dot{\gamma}(\rho_i+\rho_j)^3\right)+c_p\left(\frac{2k_BT(\rho_i+\rho_j)}{3\eta_0}(1/\rho_i+1/\rho_j)\right),$$

-------------------(2)

where γ is the effective shear rate, ρ is the aggregate radius, c is the collision efficiency for the orthokinetic (perikinetic) aggregation, T is the temperature (here T is fixed to 300 K), and η is the viscosity of the suspending fluid. The collision efficiency terms correct the aggregation kernel, which are derived for neutral impermeable particles, which in practice overestimates the collision frequency if the particles exhibit nonneutral interaction potentials or have reduced permeabilities.

Collision take place between the two erythrocytes and the aggregation dynamics is modeled in the form of mass balance equations originally written by Smoluchowski. Fragmentation is the opposite process when an aggregate splits into smaller pieces. It may happen spontaneously, due to thermal fluctuations. The fragmentation is also possible due to collisions of aggregates. In this case the kinetic energy of their relative motion transforms into kinetic energy and surface energy of debris. The analytical aggregation kernels for simple laminar shear (orthokinetic aggregation), and thermal motion (perikinetic aggregation) are used to model the coagulation kinetics of Brownian particles under laminar shear flow. The two aggregation mechanisms are independent over a wide range of conditions [17]. Therefore, the aggregation rate J (the subscripts i j are dropped to refer to equal-sized spheres) in a system undergoing both shear-induced aggregation and Brownian aggregation can be written as

$$J = \frac{8k_B T n^2}{3\eta W_B}\left[1 + \frac{4}{3\pi}\frac{W_B}{W_S}P_e\right] \text{--(3)}$$

W_B is the stability ratio introduced into the Brownian aggregation due to the potential barrier in DLVO theory, η is the plasma viscosity, γ is the surface tension and *Pe* is referred to as Péclet number, i.e.

$$Pe = \frac{3\pi\eta a^3\gamma}{k_B T} \text{---(4)}$$

W_B is the stability ratio due to dispersion forces (van der Waals-Hamaker attraction), W_S is the stability ratio which includes hydrodynamic interaction. W_B is given in terms of maximum potential barrier and stability ratio due to dispersion forces.

$$W_B = W_\infty + 0.25\left[\exp\left(\frac{\varphi_{\max}}{k_B T}\right) - 1\right] \text{--- (5)}$$

For erythrocyte spheres of identical composition in plasma where the separation distance is much smaller than the particle radius, the van der Waals contribution to the interaction potential can be calculated, to close approximation, as

$$\varphi_{vdW} = -\frac{aA_H}{12D} \text{--(6)}$$

In the limit of constant surface charge density *Φe* is approximated as

$$\varphi_e \approx 2\pi\varepsilon\varepsilon_0 a\psi_s^{\ 2} In\left[\frac{1}{1-\exp(-\kappa D)}\right] \text{--- (7)}$$

where ε is the dielectric constant of the medium, ε_0 is the permittivity of free space, ψ_S is the particle surface potential, and κ is the Debye parameter. Replacing the surface potential with the more accessible zeta potential, ζ, the total interaction potential is given by

$$\varphi = -\frac{aA_H}{12D} + 2\pi\varepsilon\varepsilon_0 a\zeta^2 In\left[\frac{1}{1-\exp(-\kappa D)}\right] \text{--(8)}$$

The bulk property yield stress, τ_y, is proportional to the inter-particle force at the separation distance times the number of particle contacts per unit area. The number of particle contacts scales as ϕ^2 / a^2 where φ is the volume fraction of the solids and a is the particle radius.

$$\tau_y \propto -\frac{\phi^2}{a^2}\frac{\varphi(D_0)}{D_0} \text{--(9)}$$

Using the interaction potential from DLVO theory and electrostatic potential, *Φe* given in equations 4 and5 and putting in Equation 7, yield stress is given by

$$\tau_y \propto \frac{\phi^2}{a}\left(\frac{A_H}{12D_0^{\ 2}} - \frac{2\pi\varepsilon\varepsilon_0}{D_0}\zeta^2 In\left[\frac{1}{1-\exp(-\kappa D_0)}\right]\right) \text{--(10)}$$

The above equation shows that at low zeta potential, van der Waals forces dominate. As the zeta potential becomes significant, yield stress increases as the square of zeta potential.
The effective volume fraction of the erythrocyte flocs is given by following equation

$$\phi_f = \left(\frac{d_f}{d_0}\right)^{3-D} \phi_0 \text{---(11)}$$

where d_f, d_0 and ϕ_0 are the diameter of a floc, diameter of a primary particle and volume fraction of primary particles. D is the fractal dimension and is in the range of 1≤ D≤3. When van der Waal forces dominate, yield stress is dependent on the diameter of the floc as

$$\tau_y = \frac{A}{9\pi d_0 h^2}\left(\frac{\phi_0}{0.52}\right)^{\frac{2}{3-D}} \text{---(12)}$$

where A is Hamaker constant and h is the minimum separation distance between particle surfaces. At the condition of maximum cubic packing, the volume fraction is 0.52. In conclusion, aggregation of erythrocytes increase as particle size decreases, hematocrit level increases, viscosity increases and shear rate decreases. Experimental results on the single cell level and theoretical analyses of the depletion model indicate the influence of the cell shape on the binding strength.

Experimental results indicate, aggregation is observed to be a two-step process: the formation of short linear stacks with few erythrocytes followed by their coalescence into long linear and branched rouleaux. As the shear rate increases the large rouleaux break up into smaller ones, and at higher values the suspension ultimately becomes one of mono-dispersed erythrocytes. This process then reverses as the shear rate is decreased [18].

Yield stress of blood is found to lie between 0.0015and 0.005 Pa at hematocrit concentratio, Ht = 0.45 i.e. a threshold stress for flow to begin often estimated by the extrapolation of measured shear stress to the zero shear rate on the basis of the Casson's equation[19]. Yield stress is power law dependent on hematocrit level and is fitted to a power law dependence and a constant that is a function of function of fibrinogen concentration. Red blood cell (RBC) agglutination depends on intermolecular attractive forces (hydrophobic bonds, Van der Wall forces, electrostatic forces and hydrogen bonds) and repulsive interactions (zeta potential). The force involved in RBC aggregation can be measured using double optical tweezers, in normal serum, in the presence of erythrocyte antibodies and associated to agglutination potentiator solutions Dextran, low ionic strength solution [LISS] and enzymes [20]. Electrostatic repulsive force extends from the surfaces of normal RBC's for at least 16 nm, preventing the aggregation of normally charged cells by the short Dextran 20. Relation between aggregating force (of fibrinogen) and disaggregating force (due to electrostatic repulsion among erythrocytes) in erythrocyte aggregation can be investigated using a rheoscope where erythrocyte aggregation increases with the increase of molecular weight of bridging macromolecules. Increase in electrostatic repulsion between cells may explain the disaggregation of normal RBC's at high concentrations of dextrans.

References:

[1] Varga-Szabo D, Pleines I, Nieswandt B. 2008. Cell adhesion mechanisms in platelets. Arterioscler Thromb Vasc Biol 28: 403–412.

[2] Doggett TA, Girdhar G, Lawshe A, Schmidtke DW, Laurenzi IJ, Diamond SL, Diacovo TG. 2002 Selectin-like kinetics and biomechanics promote rapid platelet adhesion in flow: the GPIb(alpha)-vWF tether bond. *Biophys. J.* **83**, 194–205.

[3} Tokarev, A Butylin ,A. and F. I. Ataullakhanov. 2011. Platelet adhesion from shear blood flow is controlled by near-wall rebounding collisions with erythrocytes. Biophys. J. 100:799–808.

[4] Yago T *et al.* 2008 Platelet glycoprotein Ibalpha forms catch bonds with human WT vWF but not with type 2B vonWillebrand disease vWF. *J. Clin. Invest.* **118**, 3195–3207.

[5] Ruggeri, Z. M., and G. L. Mendolicchio. 2007. Adhesion mechanisms in platelet function. Circ. Res. 100:1673–1685.

[6] Bas De Laat Clinical Advances in Hematology & Oncology Volume 12, Issue 10 October 2014.

[7] Doddi SK, Bagchi P. 2009 Three-dimensional computational modeling of multiple deformable cells flowing in microvessels. Phys. Rev. E Stat. Non lin. Soft Matter Phys. 79, 046318

[8] Zhang J, Johnson PC, Popel AS. 2007 An immersed boundary lattice Boltzmann approach to simulate deformable liquid capsules and its application to microscopic blood flows. Phys. Biol. 4, 285–295.

[9] Babler, M. U., M. Morbidelli.2007 Analysis of the aggregation fragmentation population balance equation with application to coagulation. J. Colloid Interface Sci. 316(2):428– 441,

[10] Doddi SK, Bagchi P. 2009 Three-dimensional computational modeling of multiple deformable cells flowing in microvessels. Phys. Rev. E Stat. Nonlin. Soft Matter Phys. 79, 046318. (doi:10.1103/ PhysRevE.79.046318)

[11]. A. Kumar and M. D. Graham, "Margination and seg- regation in confined flows of blood and other multicomponent suspensions," Soft Matter (2012), doi: 10.1039/C2SM25943E.

[12] O.Baskurt, B. Neu, H. Meiselman, Red Blood Cell Aggregation, CRC Press Taylor and Francis Group, 2012

[13] N. V. Brilliantov, A. S. Bodrova and P. L. Krapivsky,2009. J. Stat. Mech. P06011

[14] R. Hellmuth,1,2 M. S. Bruzzi, and N J. Quinlan, 2015,Analysis of Shear-Induced Platelet Aggregation and Breakup, Annals of Biomedical Engineering, DOI: 10.1007/s10439-015-1409-1

[15]P. Decuzzi and M. Ferrari, "Modulating cellular adhesion through nanotopography," Biomaterials, vol. 31, no. 1, pp. 173–179, 2010.

[16] J.Jung, 1, Byoung-Kwon Lee, and S Shin.2014, Yield shear stress and disaggregating shear stress of human blood, Korea-Australia Rheology Journal, Vol.26, No.2,.191-198
.[17]M Mohtaschemi, A Puisto, X Illab and M. J. Alavaa, 2014, Rheology dynamics of aggregating colloidal suspensions, Soft Matter, 2014, 10, 2971.
[18] [17] Q. Zhao, L. G. Durand, L. Allard, and G. Cloutier, Effects of a sudden flow reduction on red blood cell rouleau formation and orientation using RF backscattered power, Ultrasound Med. Biol., **24**, 503–511, 1998
[19] A. L. Copley, C. R. Huang, and R. G. King,1973, Rheogoniometric studies of whole human blood at shear rates from 1,000-0.0009 sec−1. Part I. Experimental findings, Biorheology, **10**, 17–22.
[20] H. Bäumler and E. Donath, Does dextran realy significantly increase the surface potential of human red blood cells?, Stud.Biophysica **120** (1987), 113–122.

420230 Big Data Analysis for Selecting Clinically Relevant Biomarkers: A Global Optimization Framework

Abstract:

Biomarkers are measurable indicators of biological processes that can be applied in clinical settings for disease diagnosis and prognosis, risk-factor assessment, disease staging, and as indicators of treatment efficacy. High-throughput –omics platforms enables the interface of personalized medicine with the advantages and challenges of big data. Typical studies generate expansive datasets with thousands of candidate biomarkers (i.e., data features). For a biomarker to be accepted into clinical praxis, it must be subjected to large-scale, often expensive clinical validation stages [1]; the ultimate success of a discovery-phase biomarker study lies in its ability to produce a small subset of biomarkers with the greatest probability of success in large-scale targeted studies [1,2]. This high data dimensionality per sample is almost always coupled with a comparatively low number of samples in the discovery phase, yielding a statistically difficult feature selection problem with a high probability of overfitting or of selecting data artifacts as meaningful candidates [3].

We present a novel optimization model that selects the optimal subset of candidate biomarkers. The model is self-regularized and can be solved to global optimality to obtain the best subset of biomarkers as per the objective function. We display the utility and power of the model by applying it to four well-known expansive cancer genomics datasets from the literature [4-7] which study differences between diseased and healthy patients or between different types of cancer. All datasets have an extremely low (< 0.02) samples-to-features ratio. We utilized an established evaluation protocol [8] to compare our method to the current state of the art [9]. Very small subsets of genes selected by the model are able to classify new samples from each of the disease systems with high sensitivity, specificity, and accuracy, and are selected robustly and stably in the face of random data permutations [8,10,11]. The model comprises a general methodology which applies to any scenario, biomedical or otherwise, where an optimally descriptive set of features must be extracted from a vast pool of mostly irrelevant possibilities.

References:

1. Rifai, N., Gillette, M. A., & Carr, S. A. (2006). Protein biomarker discovery and validation: the long and uncertain path to clinical utility. *Nature biotechnology*, *24*(8), 971-983.

2. Srinivas, P. R., Verma, M., Zhao, Y., & Srivastava, S. (2002). Proteomics for cancer biomarker discovery. *Clinical chemistry*, *48*(8), 1160-1169.

3. Rubingh, C. M., Bijlsma, S., Derks, E. P., Bobeldijk, I., Verheij, E. R., Kochhar, S., & Smilde, A. K. (2006). Assessing the performance of statistical validation tools for megavariate metabolomics data. *Metabolomics*, *2*(2), 53-61.

4. Alon, U., Barkai, N., Notterman, D. A., Gish, K., Ybarra, S., Mack, D., & Levine, A. J. (1999). Broad patterns of gene expression revealed by clustering analysis of tumor and normal colon tissues probed by oligonucleotide arrays. *Proceedings of the National Academy of Sciences*, *96*(12), 6745-6750.

5. Golub, T. R., Slonim, D. K., Tamayo, P., Huard, C., Gaasenbeek, M., Mesirov, J. P., ... & Lander, E. S. (1999). Molecular classification of cancer: class discovery and class prediction by gene expression monitoring. *science*, *286*(5439), 531-537.

6. Alizadeh, A. A., Eisen, M. B., Davis, R. E., Ma, C., Lossos, I. S., Rosenwald, A., ... & Staudt, L. M. (2000). Distinct types of diffuse large B-cell lymphoma identified by gene expression profiling. *Nature*, *403*(6769), 503-511.

7. Singh, D., Febbo, P. G., Ross, K., Jackson, D. G., Manola, J., Ladd, C., ... & Sellers, W. R. (2002). Gene expression correlates of clinical prostate cancer behavior. *Cancer cell*, *1*(2), 203-209.

8. Abeel, T., Helleputte, T., Van de Peer, Y., Dupont, P., & Saeys, Y. (2010). Robust biomarker identification for cancer diagnosis with ensemble feature selection methods. *Bioinformatics*, *26*(3), 392-398.

9. Guyon, I., Weston, J., Barnhill, S., & Vapnik, V. (2002). Gene selection for cancer classification using support vector machines. *Machine learning*, *46*(1-3), 389-422.

10. He, Z., & Yu, W. (2010). Stable feature selection for biomarker discovery. *Computational biology and chemistry*, *34*(4), 215-225.

11. Kuncheva, L. I. (2007, February). A stability index for feature selection. In *Artificial intelligence and applications* (pp. 421-427).

Copyright © American Institute of Chemical Engineers. All rights reserved.

AUTHOR INDEX

AUTHOR INDEX